1996 YEARBOOK of ASTRONOMY

1996 YEARBOOK of ASTRONOMY

edited by

Patrick Moore

MACMILLAN

First published 1995 by
Macmillan Reference Books
a division of Macmillan Publishers Limited
25 Eccleston Place, London SW1W 9NF
and Basingstoke

Associated companies throughout the world

ISBN 0–333–63702–X

9 8 7 6 5 4 3 2 1

A CIP catalogue record for this book is available from the British Library

Photoset by Rowland Phototypesetting Limited
Bury St Edmunds, Suffolk
Printed and bound in Great Britain by
Mackays of Chatham plc, Chatham, Kent

Contents

Part II: *Article Section*

Part III: *Miscellaneous*

Editor's Foreword

This edition of the *Yearbook* follows the familiar pattern; we have our usual monthly notes, compiled by Gordon Taylor, together with the familiar lists and our selection of articles. We welcome new authors as well as those who have contributed so often before.

There is, however, one difference. For many years the opening article has been written by David Allen, who was not only one of our most brilliant astronomers of the 'younger' school, but was also a particularly good writer of general articles – and a loyal friend as well. Sadly, as readers of last year's *Yearbook* will know, David Allen has lost his long and brave battle against cancer. He will be greatly missed by everyone; he will not be forgotten. This issue of the *Yearbook* is dedicated to his memory.

PATRICK MOORE
Selsey, May 1995

Preface

New readers will find that all the information in this *Yearbook* is given in diagrammatic or descriptive form; the positions of the planets may easily be found on the specially designed star charts, while the monthly notes describe the movements of the planets and give details of other astronomical phenomena visible in both the northern and southern hemispheres. Two sets of the star charts are provided. The **Northern Charts** (pp. 14 to 39) are designed for use in latitude 52 degrees north, but may be used without alteration throughout the British Isles, and (except in the case of eclipses and occultations) in other countries of similar north latitude. The **Southern Charts** (pp. 40 to 65) are drawn for latitude 35 degrees south, and are suitable for use in South Africa, Australia and New Zealand, and other stations in approximately the same south latitude. The reader who needs more detailed information will find *Norton's Star Atlas* (Longman) an invaluable guide, while more precise positions of the planets and their satellites, together with predictions of occultations, meteor showers, and periodic comets may be found in the *Handbook* of the British Astronomical Association. The British monthly periodical, with current news, articles, and monthly notes is *Astronomy Now*. Readers will also find details of forthcoming events given in the American *Sky and Telescope*. This monthly publication also produces a special occultation supplement giving predictions for the United States and Canada.

Important Note
The times given on the star charts and in the Monthly Notes are generally given as local times, using the 24-hour clock, the day beginning at midnight. All the dates, and the times of a few events (e.g. eclipses), are given in Greenwich Mean Time (GMT), which is related to local time by the formula

Local Mean Time = GMT − west longitude

In practice, small differences of longitudes are ignored, and the observer will use local clock time, which will be the appropriate

Standard (or Zone) Time. As the formula indicates, places in west longitude will have a Standard Time slow on GMT, while places in east longitude will have a Standard Time fast on GMT. As examples we have:

Standard Time in

New Zealand	GMT	+	12 hours
Victoria; N.S.W.	GMT	+	10 hours
Western Australia	GMT	+	8 hours
South Africa	GMT	+	2 hours
British Isles	GMT		
Eastern S.T.	GMT	−	5 hours
Central S.T.	GMT	−	6 hours, etc.

If Summer Time is in use, the clocks will have to have been advanced by one hour, and this hour must be subtracted from the clock time to give Standard Time.

In Great Britain and N. Ireland, Summer Time will be in force in 1996 from March 31 until October 27 GMT.

Notes on the Star Charts

The stars, together with the Sun, Moon and planets seem to be set on the surface of the celestial sphere, which appears to rotate about the Earth from east to west. Since it is impossible to represent a curved surface accurately on a plane, any kind of star map is bound to contain some form of distortion. But it is well known that the eye can endure some kinds of distortion better than others, and it is particularly true that the eye is most sensitive to deviations from the vertical and horizontal. For this reason the star charts given in this volume have been designed to give a true representation of vertical and horizontal lines, whatever may be the resulting distortion in the shape of a constellation figure. It will be found that the amount of distortion is, in general, quite small, and is only obvious in the case of large constellations such as Leo and Pegasus, when these appear at the top of the charts, and so are drawn out sideways.

The charts show all stars down to the fourth magnitude, together with a number of fainter stars which are necessary to define the shape of a constellation. There is no standard system for representing the outlines of the constellations, and triangles and other simple figures have been used to give outlines which are easy to follow with the naked eye. The names of the constellations are given, together with the proper names of the brighter stars. The apparent magnitudes of the stars are indicated roughly by using four different sizes of dots, the larger dots representing the brighter stars.

The two sets of star charts are similar in design. At each opening there is a group of four charts which give a complete coverage of the sky up to an altitude of 62½ degrees; there are twelve such groups to cover the entire year. In the **Northern Charts** (for 52 degrees north) the upper two charts show the southern sky, south being at the centre and east on the left. The coverage is from 10 degrees north of east (top left) to 10 degrees north of west (top right). The two lower charts show the northern sky from 10 degrees south of west (lower left) to 10 degrees south of east (lower right). There is thus an overlap east and west.

Conversely, in the **Southern Charts** (for 35 degrees south) the upper two charts show the northern sky, with north at the centre and east on the right. The two lower charts show the southern sky, with south at the centre and east on the left. The coverage and overlap is the same on both sets of charts.

Because the sidereal day is shorter than the solar day, the stars appear to rise and set about four minutes earlier each day, and this amounts to two hours in a month. Hence the twelve groups of charts in each set are sufficient to give the appearance of the sky throughout the day at intervals of two hours, or at the same time of night at monthly intervals throughout the year. The actual range of dates and times when the stars on the charts are visible is indicated at the top of each page. Each group is numbered in bold type, and the number to be used for any given month and time is summarized in the following table:

Local Time	18h	20h	22h	0h	2h	4h	6h
January	11	12	1	2	3	4	5
February	12	1	2	3	4	5	6
March	1	2	3	4	5	6	7
April	2	3	4	5	6	7	8
May	3	4	5	6	7	8	9
June	4	5	6	7	8	9	10
July	5	6	7	8	9	10	11
August	6	7	8	9	10	11	12
September	7	8	9	10	11	12	1
October	8	9	10	11	12	1	2
November	9	10	11	12	1	2	3
December	10	11	12	1	2	3	4

The charts are drawn to scale, the horizontal measurements, marked at every 10 degrees, giving the azimuths (or true bearings) measured from the north round through east (90 degrees), south (180 degrees), and west (270 degrees). The vertical measurements, similarly marked, give the altitudes of the stars up to 62½ degrees. Estimates of altitude and azimuth made from these charts will necessarily be mere approximations, since no observer will be exactly at the adopted latitude, or at the stated time, but they will serve for the identification of stars and planets.

The ecliptic is drawn as a broken line on which longitude is marked at every 10 degrees; the positions of the planets are then easily found by reference to the table on page 71. It will be noticed that on the Southern Charts the **ecliptic** may reach an altitude in excess of 62½ degrees on star charts 5 to 9. The continuations of the broken line will be found on the charts of overhead stars.

There is a curious illusion that stars at an altitude of 60 degrees or more are actually overhead, and the beginner may often feel that he is leaning over backwards in trying to see them. These overhead stars are given separately on the pages immediately following the main star charts. The entire year is covered at one opening, each of the four maps showing the overhead stars at times which correspond to those of three of the main star charts. The position of the zenith is indicated by a cross, and this cross marks the centre of a circle which is 35 degrees from the zenith; there is thus a small overlap with the main charts.

The broken line leading from the north (on the Northern Charts) or from the south (on the Southern Charts) is numbered to indicate the corresponding main chart. Thus on page 38 the N-S line numbered 6 is to be regarded as an extension of the centre (south) line of chart 6 on pages 24 and 25, and at the top of these pages are printed the dates and times which are appropriate. Similarly, on page 65, the S-N line numbered 10 connects with the north line of the upper charts on pages 58 and 59.

The overhead stars are plotted as maps on a conical projection, and the scale is rather smaller than that of the main charts.

1L

October 6 at 5ʰ	October 21 at 4ʰ
November 6 at 3ʰ	November 21 at 2ʰ
December 6 at 1ʰ	December 21 at midnight
January 6 at 23ʰ	January 21 at 22ʰ
February 6 at 21ʰ	February 21 at 20ʰ

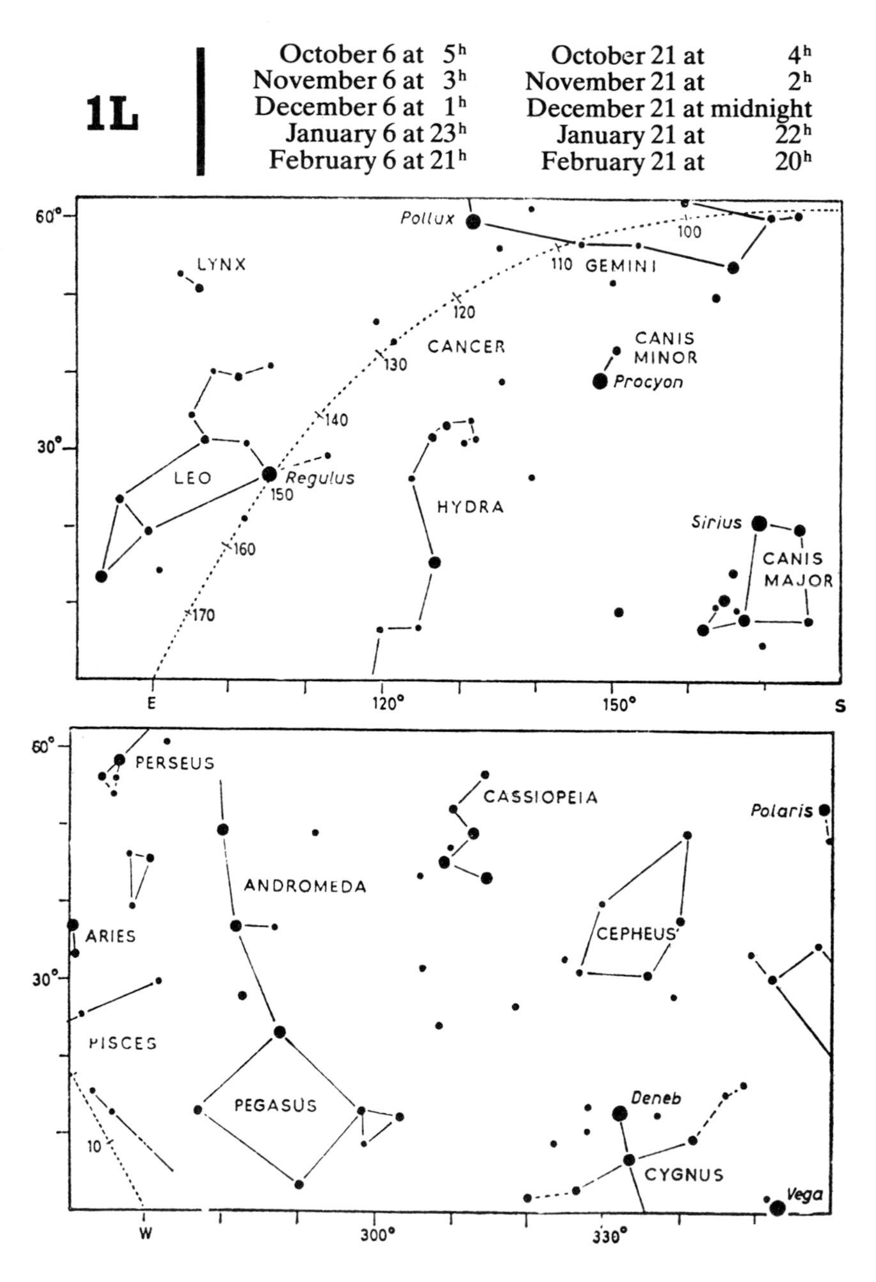

October 6 at 5^h	October 21 at 4^h	**1R**
November 6 at 3^h	November 21 at 2^h	
December 6 at 1^h	December 21 at midnight	
January 6 at 23^h	January 21 at 22^h	
February 6 at 21^h	February 21 at 20^h	

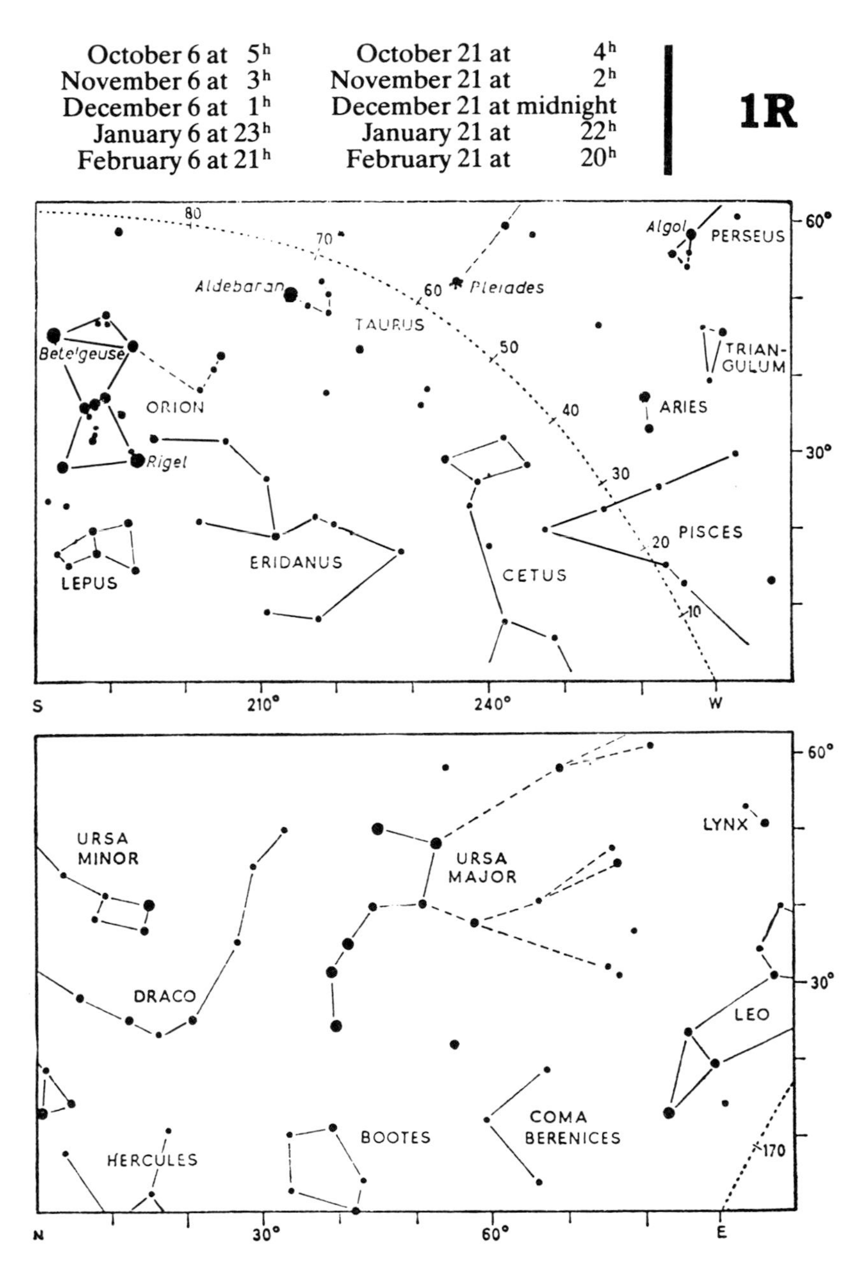

2L

November 6 at 5^h	November 21 at 4^h
December 6 at 3^h	December 21 at 2^h
January 6 at 1^h	January 21 at midnight
February 6 at 23^h	February 21 at 22^h
March 6 at 21^h	March 21 at 20^h

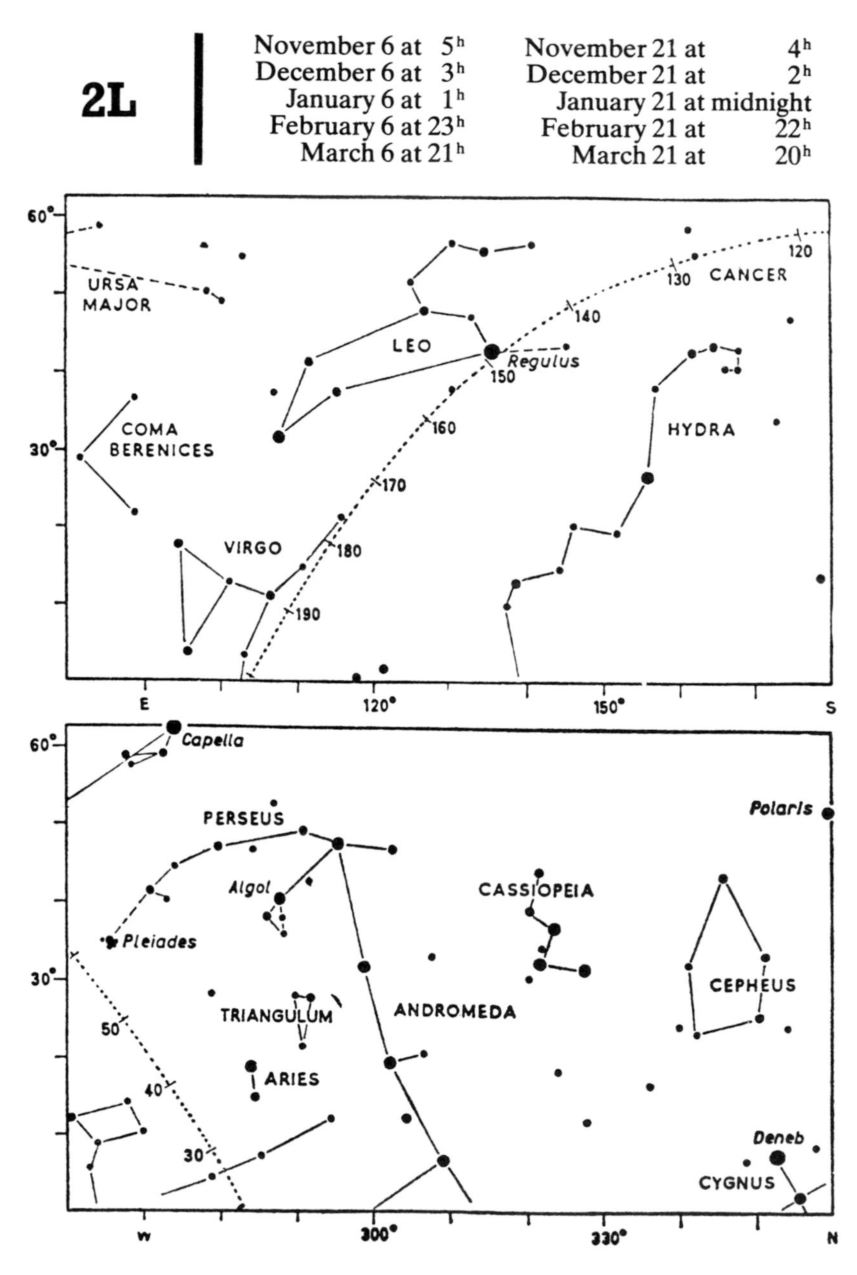

November 6 at 5ʰ	November 21 at 4ʰ
December 6 at 3ʰ	December 21 at 2ʰ
January 6 at 1ʰ	January 21 at midnight
February 6 at 23ʰ	February 21 at 22ʰ
March 6 at 21ʰ	March 21 at 20ʰ

2R

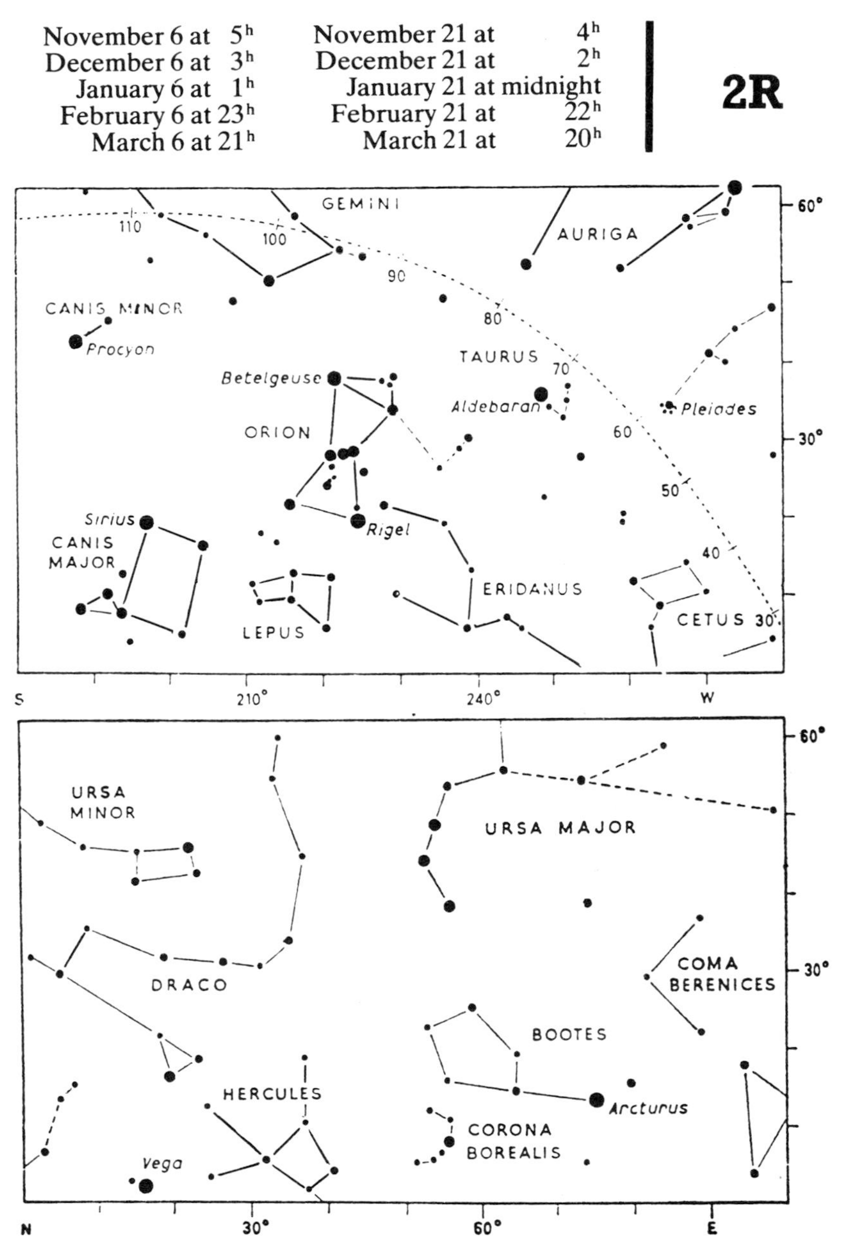

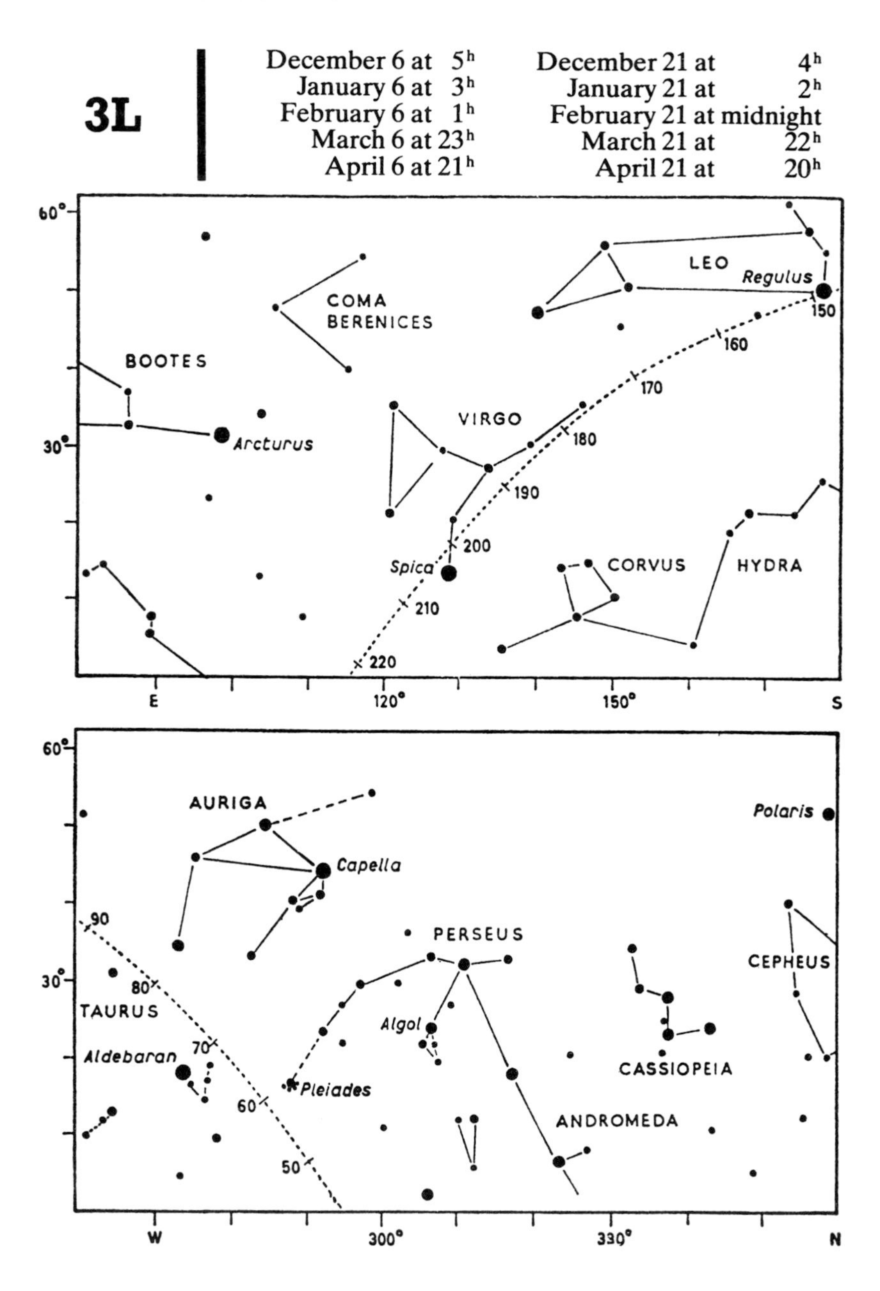
3L
December 6 at 5h
January 6 at 3h
February 6 at 1h
March 6 at 23h
April 6 at 21h
December 21 at 4h
January 21 at 2h
February 21 at midnight
March 21 at 22h
April 21 at 20h
60°
30°
BOOTES
Arcturus
COMA
BERENICES
VIRGO
Spica
LEO
Regulus
CORVUS
HYDRA
150
160
170
180
190
200
210
220
E
120°
150°
S
60°
30°
AURIGA
Capella
Polaris
PERSEUS
CEPHEUS
TAURUS
Aldebaran
Algol
Pleiades
CASSIOPEIA
ANDROMEDA
90
80
70
60
50
W
300°
330°
N

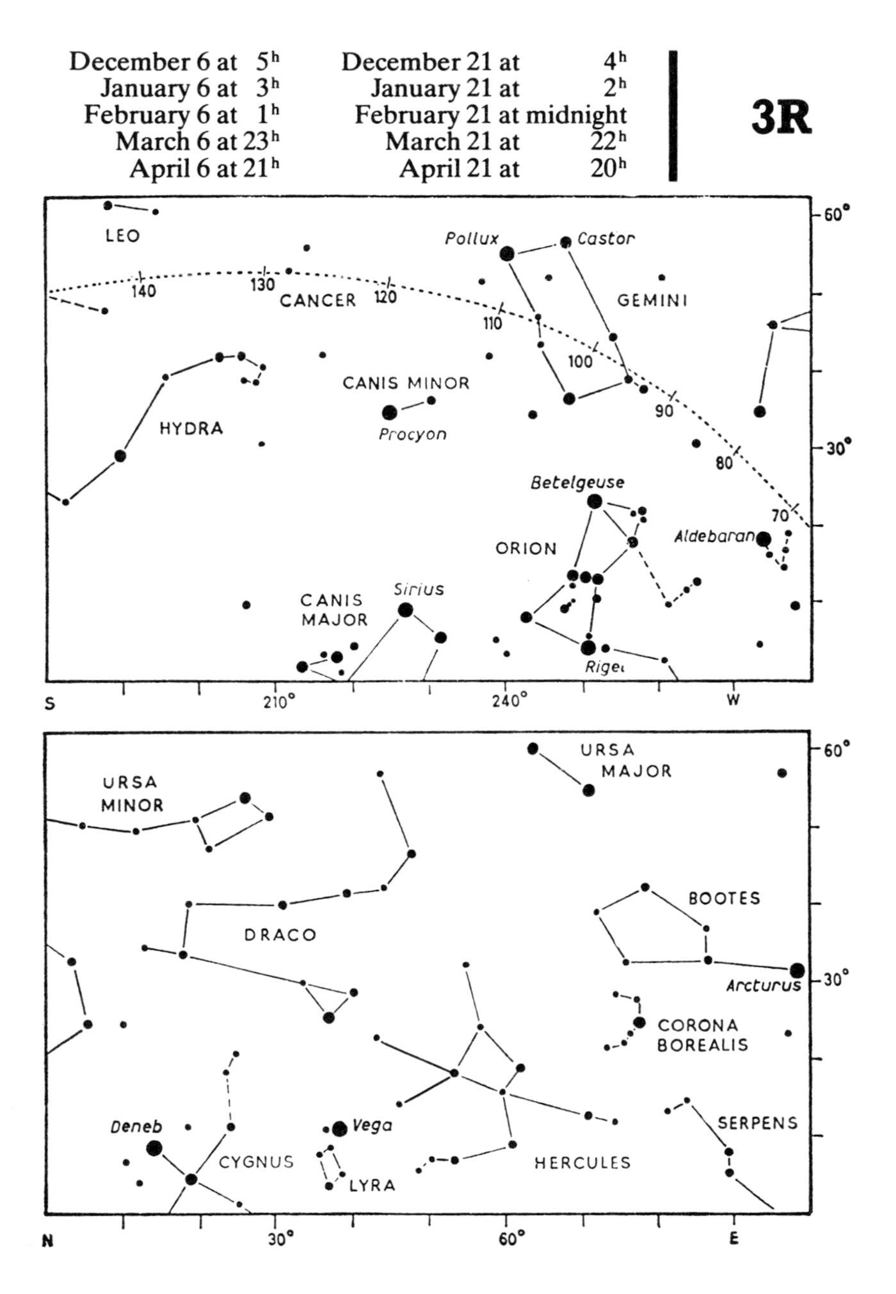
December 6 at 5h
January 6 at 3h
February 6 at 1h
March 6 at 23h
April 6 at 21h
December 21 at 4h
January 21 at 2h
February 21 at midnight
March 21 at 22h
April 21 at 20h
3R
LEO
Pollux
Castor
CANCER
GEMINI
140
130
120
110
100
90
80
70
CANIS MINOR
Procyon
HYDRA
Betelgeuse
ORION
Aldebaran
CANIS MAJOR
Sirius
Rige
60°
30°
S
210°
240°
W
URSA MAJOR
URSA MINOR
DRACO
BOOTES
Arcturus
CORONA BOREALIS
SERPENS
HERCULES
Vega
LYRA
Deneb
CYGNUS
60°
30°
N
30°
60°
E

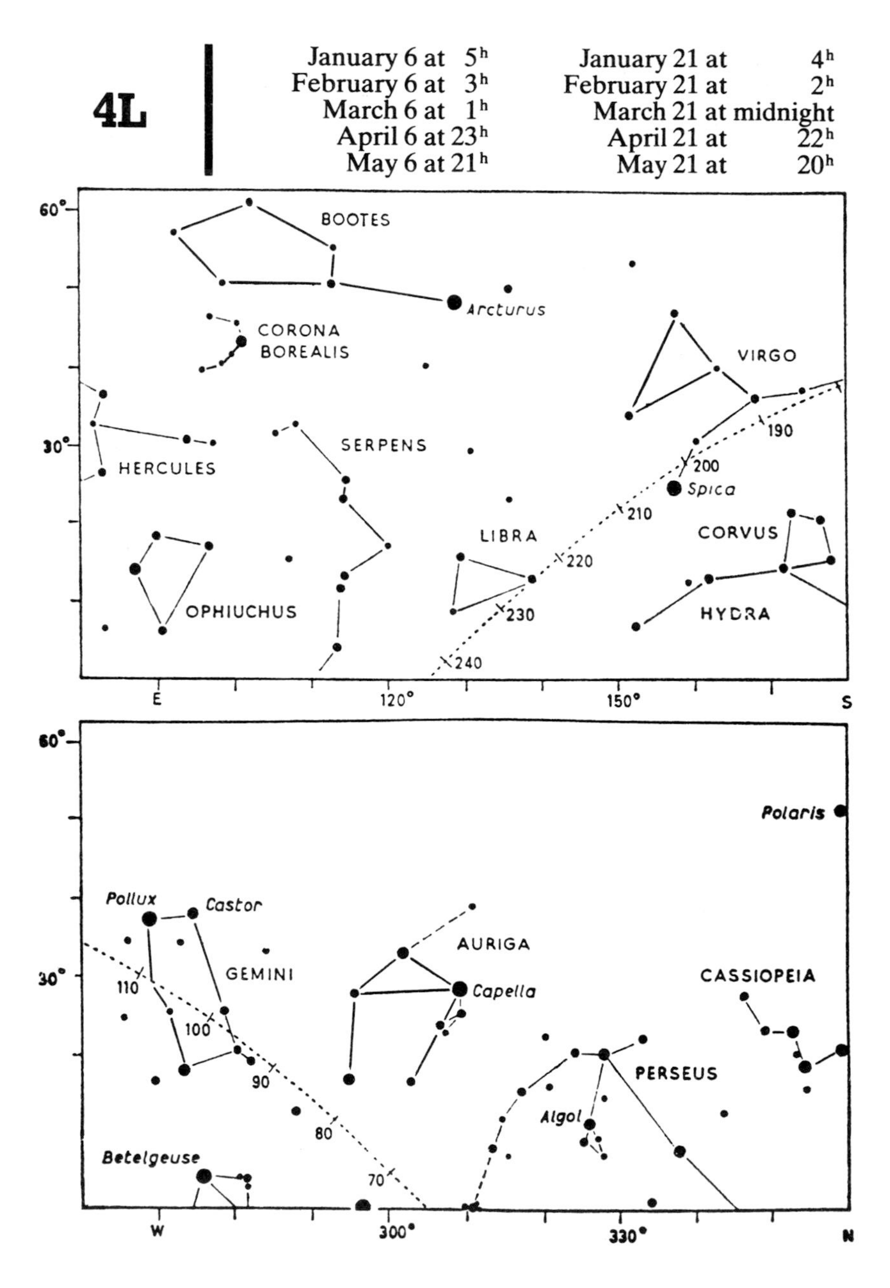
4L
January 6 at 5h
February 6 at 3h
March 6 at 1h
April 6 at 23h
May 6 at 21h
January 21 at 4h
February 21 at 2h
March 21 at midnight
April 21 at 22h
May 21 at 20h
60°
30°
BOOTES
Arcturus
CORONA BOREALIS
VIRGO
HERCULES
SERPENS
190
200
Spica
210
LIBRA
CORVUS
220
OPHIUCHUS
230
HYDRA
240
E
120°
150°
S
60°
Polaris
Pollux
Castor
AURIGA
30°
110
GEMINI
Capella
CASSIOPEIA
100
90
PERSEUS
80
Algol
Betelgeuse
70
W
300°
330°
N

4R

January 6 at 5h	January 21 at 4h
February 6 at 3h	February 21 at 2h
March 6 at 1h	March 21 at midnight
April 6 at 23h	April 21 at 22h
May 6 at 21h	May 21 at 20h

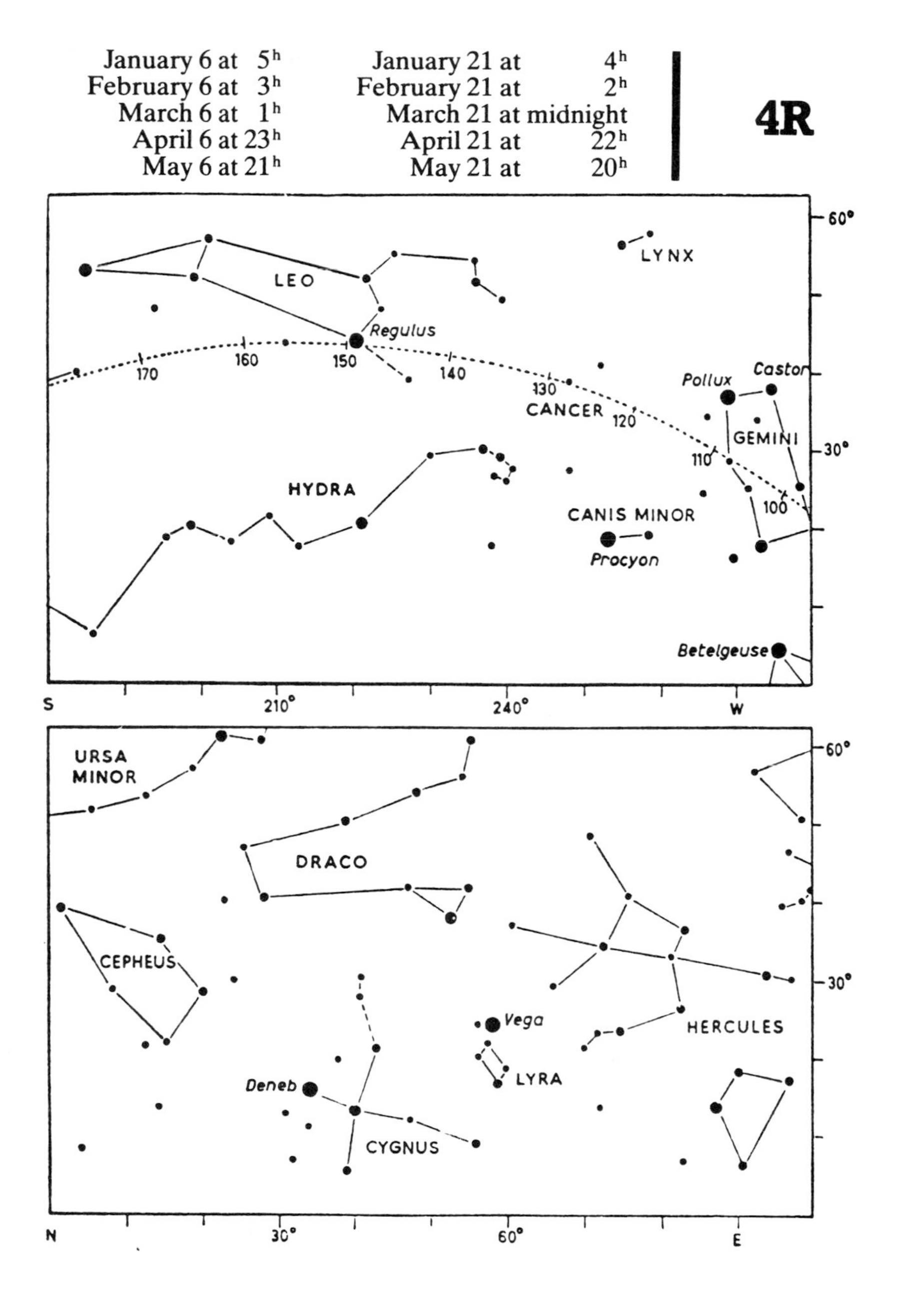

5L

January 6 at 7ʰ	January 21 at 6ʰ
February 6 at 5ʰ	February 21 at 4ʰ
March 6 at 3ʰ	March 21 at 2ʰ
April 6 at 1ʰ	April 21 at midnight
May 6 at 23ʰ	May 21 at 22ʰ

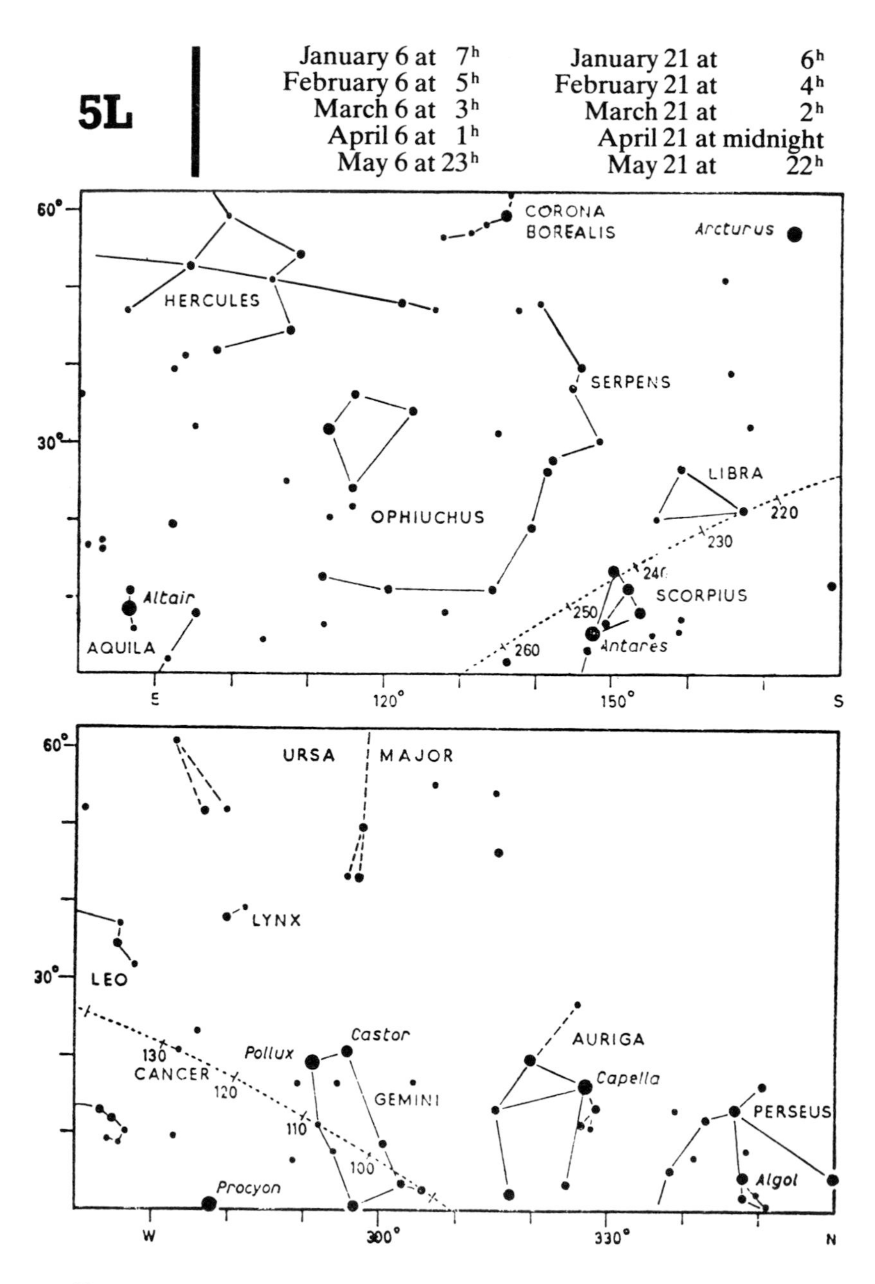

January 6 at 7^h	January 21 at 6^h		**5R**
February 6 at 5^h	February 21 at 4^h		
March 6 at 3^h	March 21 at 2^h		
April 6 at 1^h	April 21 at midnight		
May 6 at 23^h	May 21 at 22^h		

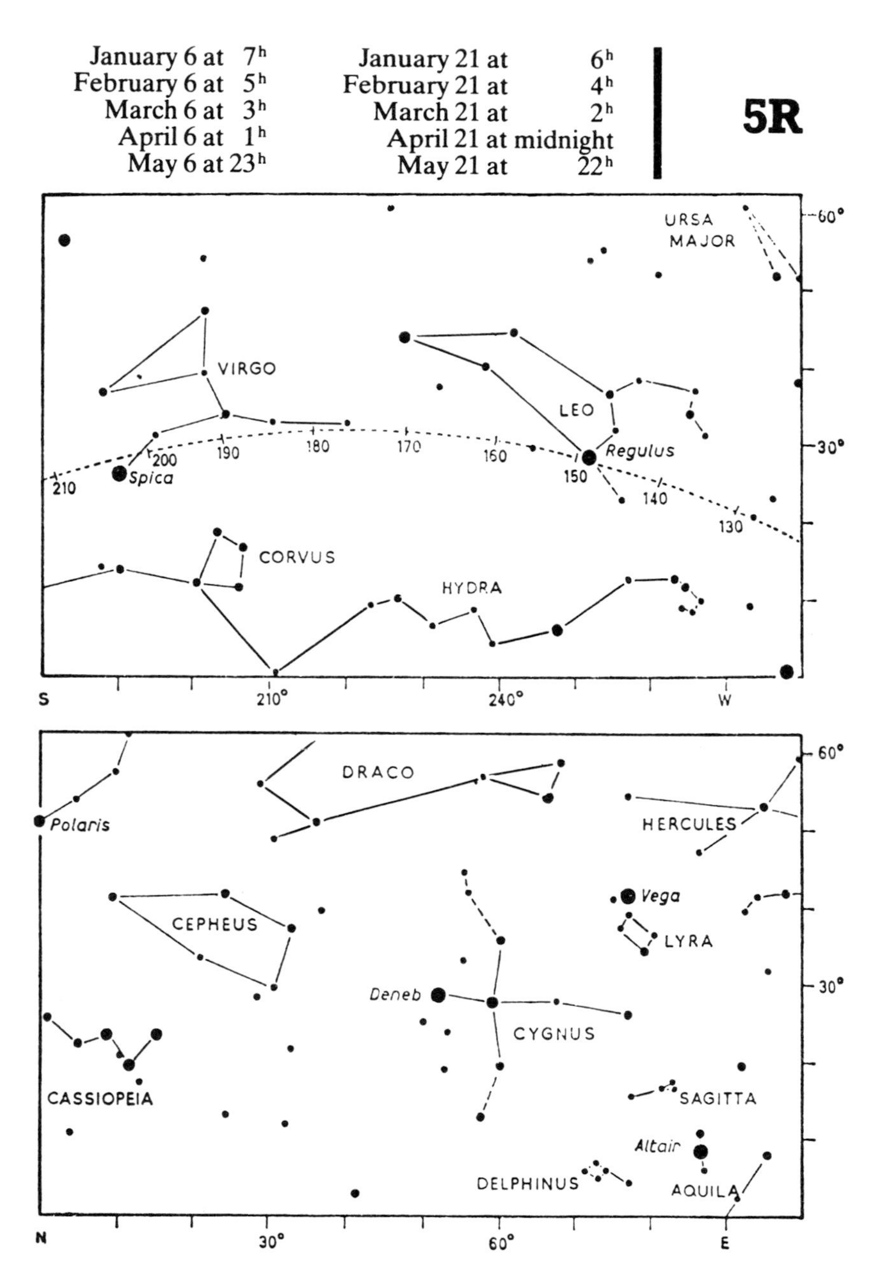

6L

March 6 at 5h	March 21 at 4h
April 6 at 3h	April 21 at 2h
May 6 at 1h	May 21 at midnight
June 6 at 23h	June 21 at 22h
July 6 at 21h	July 21 at 20h

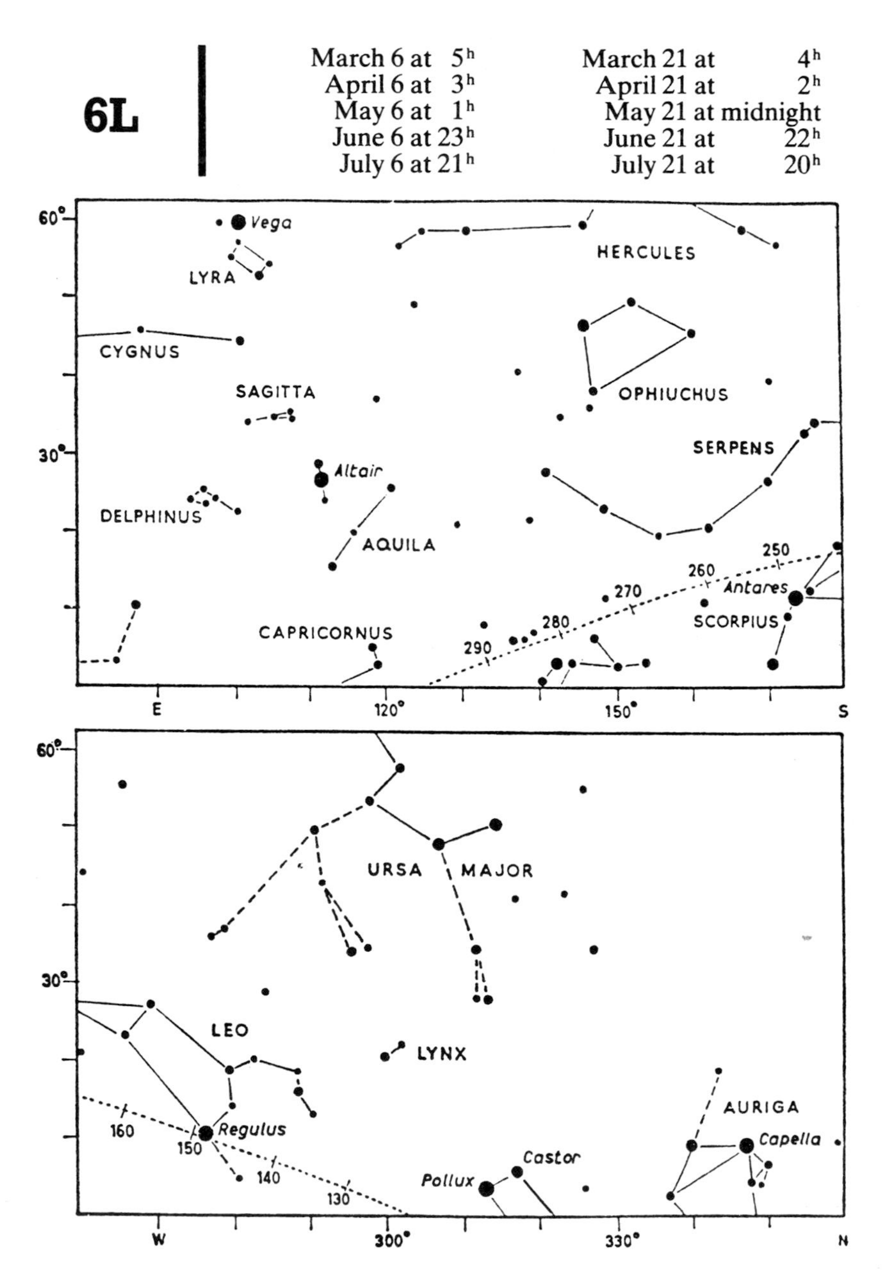

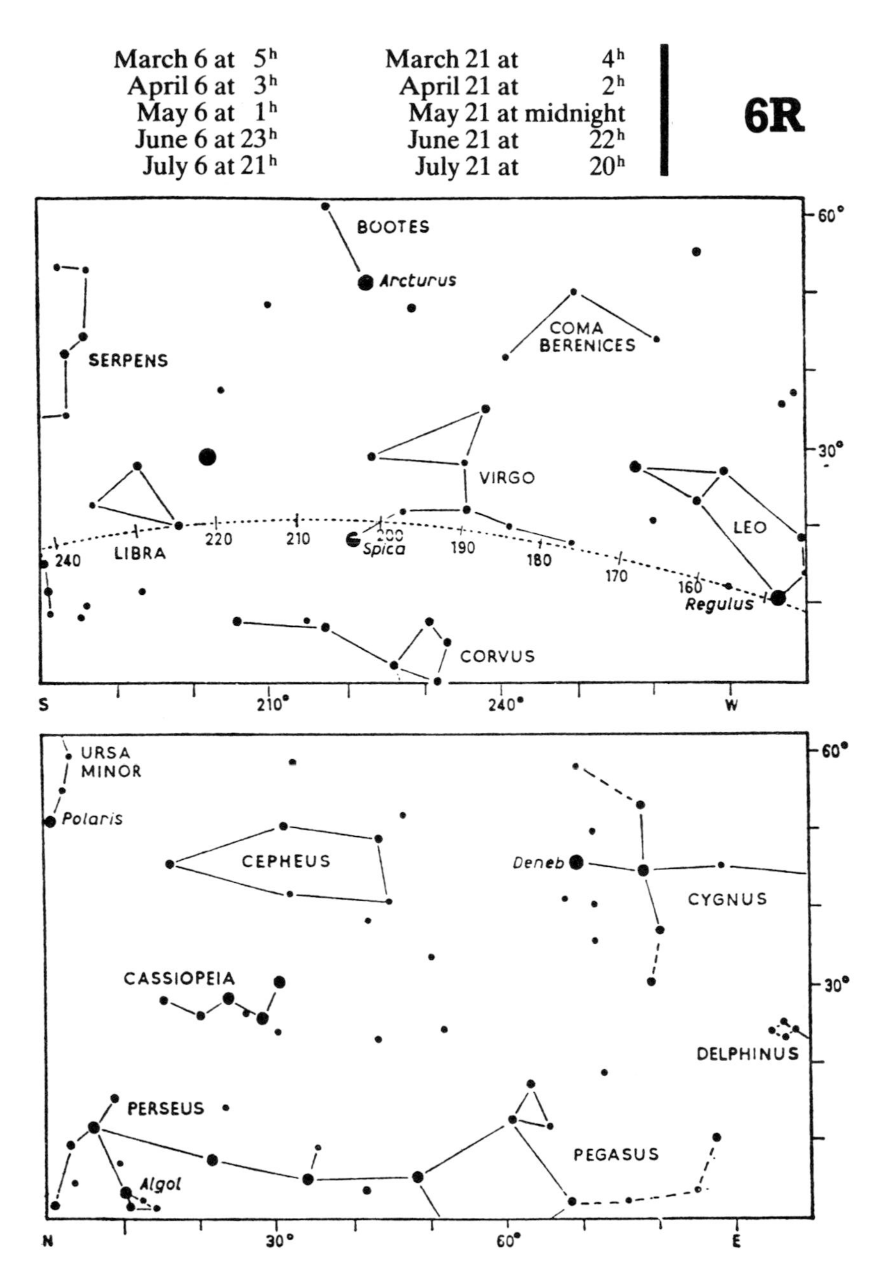
March 6 at 5^{h}
April 6 at 3^{h}
May 6 at 1^{h}
June 6 at 23^{h}
July 6 at 21^{h}
March 21 at 4^{h}
April 21 at 2^{h}
May 21 at midnight
June 21 at 22^{h}
July 21 at 20^{h}
6R
BOOTES
Arcturus
SERPENS
COMA BERENICES
VIRGO
LIBRA
LEO
Spica
Regulus
CORVUS
240
220
210
200
190
180
170
160
60°
30°
S
210°
240°
W
URSA MINOR
Polaris
CEPHEUS
Deneb
CYGNUS
CASSIOPEIA
DELPHINUS
PERSEUS
Algol
PEGASUS
60°
30°
N
30°
60°
E

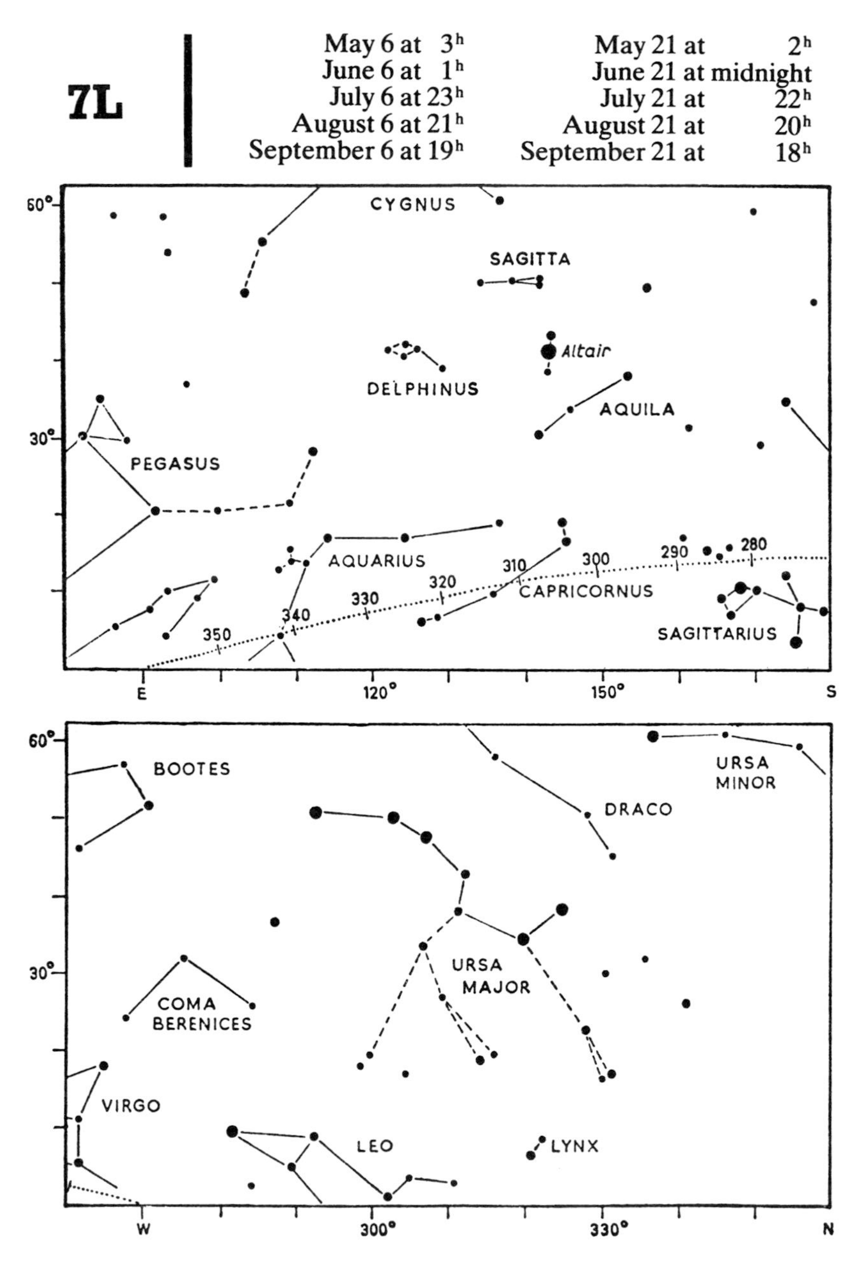
7L
May 6 at 3^{h}
June 6 at 1^{h}
July 6 at 23^{h}
August 6 at 21^{h}
September 6 at 19^{h}
May 21 at 2^{h}
June 21 at midnight
July 21 at 22^{h}
August 21 at 20^{h}
September 21 at 18^{h}
60°
30°
CYGNUS
SAGITTA
Altair
DELPHINUS
AQUILA
PEGASUS
AQUARIUS
CAPRICORNUS
SAGITTARIUS
350
340
330
320
310
300
290
280
E
120°
150°
S
60°
30°
BOOTES
URSA MINOR
DRACO
URSA MAJOR
COMA BERENICES
VIRGO
LEO
LYNX
W
300°
330°
N

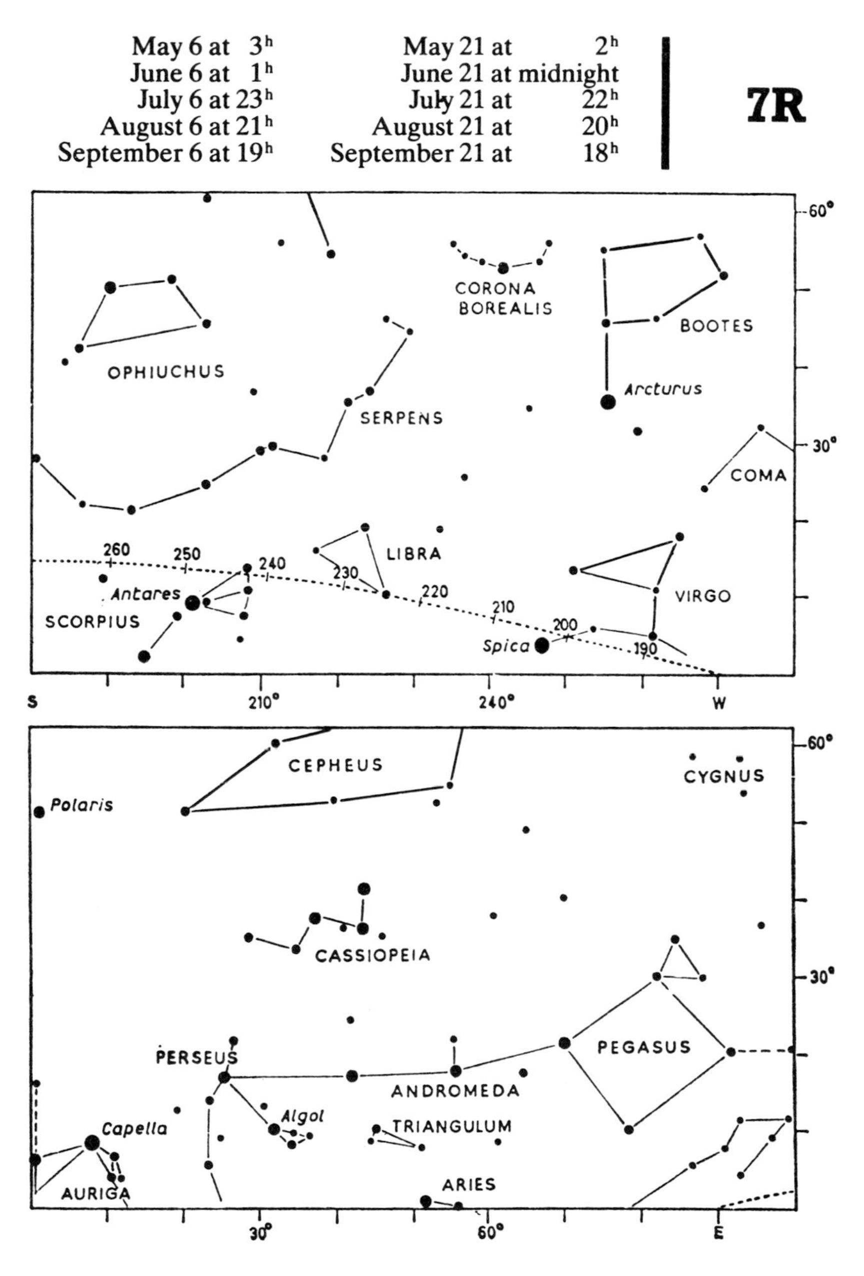
May 6 at 3h
June 6 at 1h
July 6 at 23h
August 6 at 21h
September 6 at 19h
May 21 at 2h
June 21 at midnight
July 21 at 22h
August 21 at 20h
September 21 at 18h
7R
CORONA BOREALIS
BOOTES
OPHIUCHUS
SERPENS
Arcturus
COMA
LIBRA
VIRGO
Antares
SCORPIUS
Spica
260
250
240
230
220
210
200
190
60°
30°
S
210°
240°
W
CEPHEUS
CYGNUS
Polaris
CASSIOPEIA
PEGASUS
PERSEUS
ANDROMEDA
Algol
Capella
TRIANGULUM
ARIES
AURIGA
30°
60°
E

8L

July 6 at 1^h	July 21 at midnight
August 6 at 23^h	August 21 at 22^h
September 6 at 21^h	September 21 at 20^h
October 6 at 19^h	October 21 at 18^h
November 6 at 17^h	November 21 at 16^h

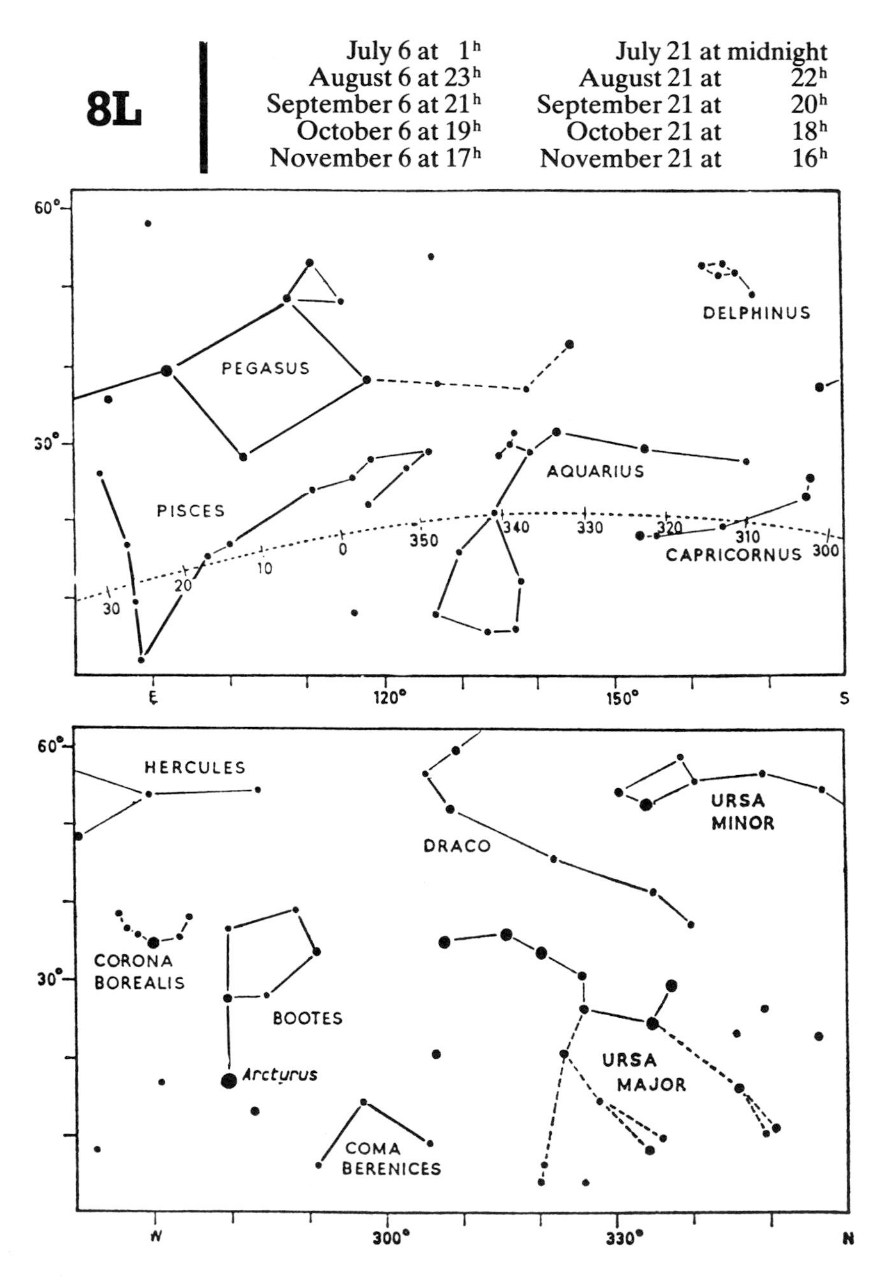

July 6 at 1^h	July 21 at midnight	**8R**
August 6 at 23^h	August 21 at 22^h	
September 6 at 21^h	September 21 at 20^h	
October 6 at 19^h	October 21 at 18^h	
November 6 at 17^h	November 21 at 16^h	

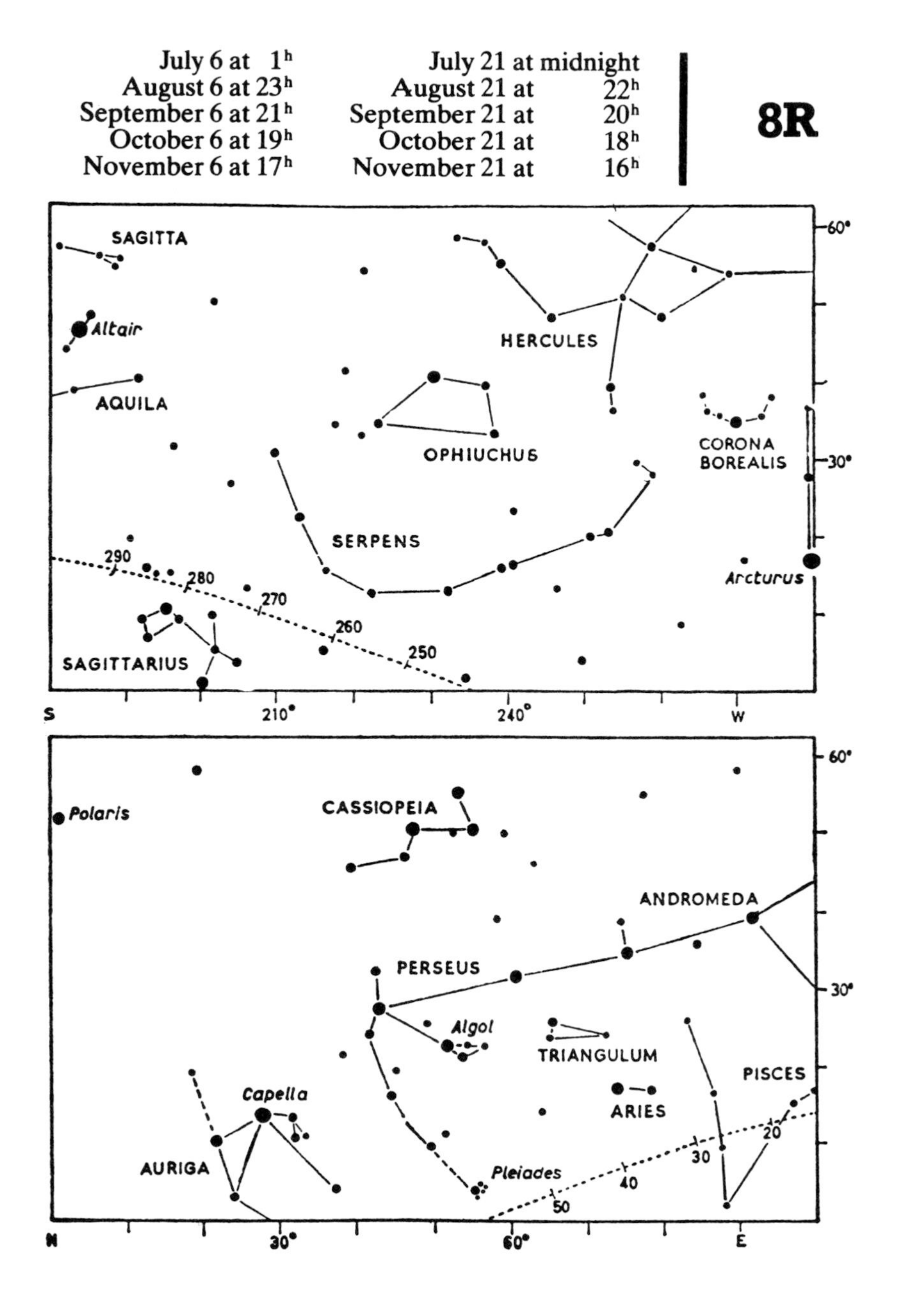

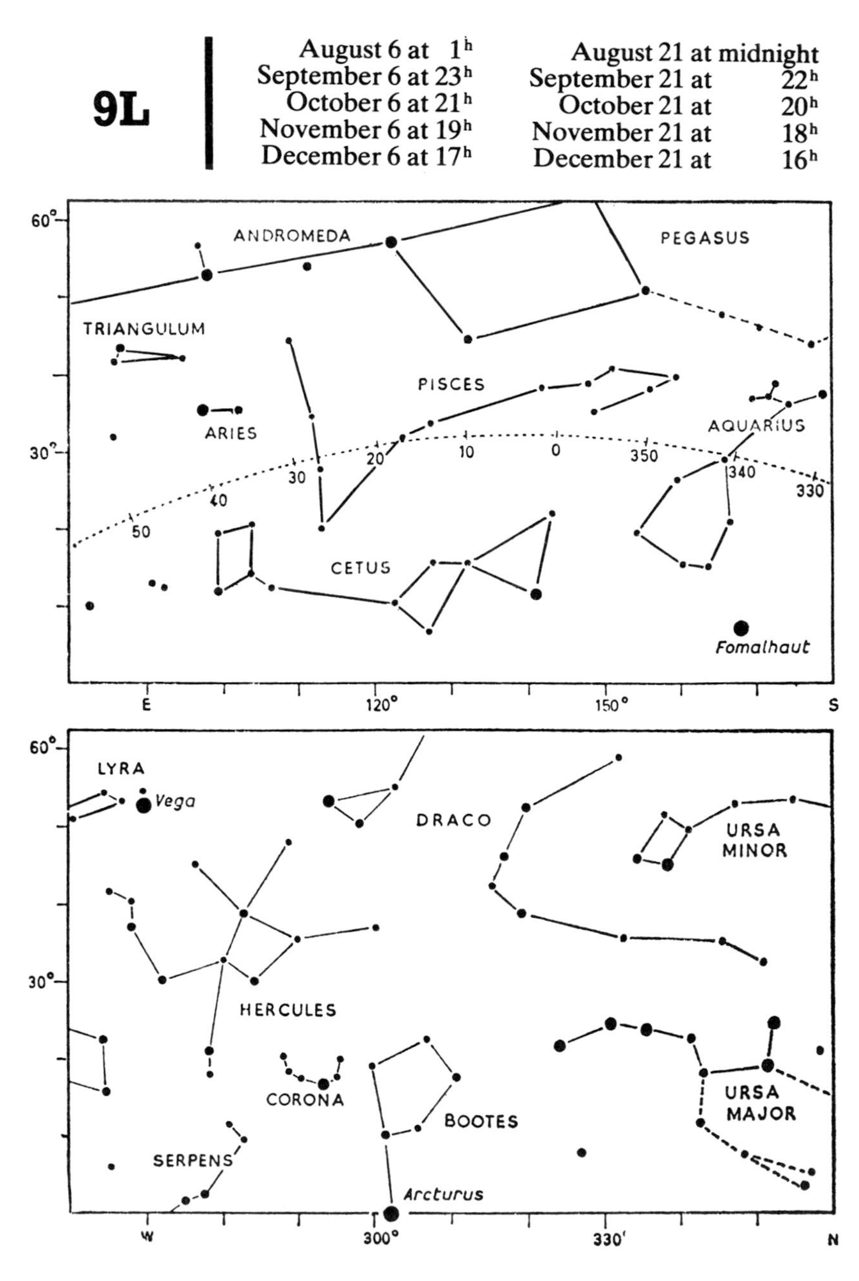
9L
August 6 at 1h
September 6 at 23h
October 6 at 21h
November 6 at 19h
December 6 at 17h
August 21 at midnight
September 21 at 22h
October 21 at 20h
November 21 at 18h
December 21 at 16h
60°
30°
ANDROMEDA
PEGASUS
TRIANGULUM
PISCES
ARIES
AQUARIUS
CETUS
Fomalhaut
50
40
30
20
10
0
350
340
330
E
120°
150°
S
60°
30°
LYRA
Vega
DRACO
URSA
MINOR
HERCULES
CORONA
BOOTES
URSA
MAJOR
SERPENS
Arcturus
W
300°
330°
N

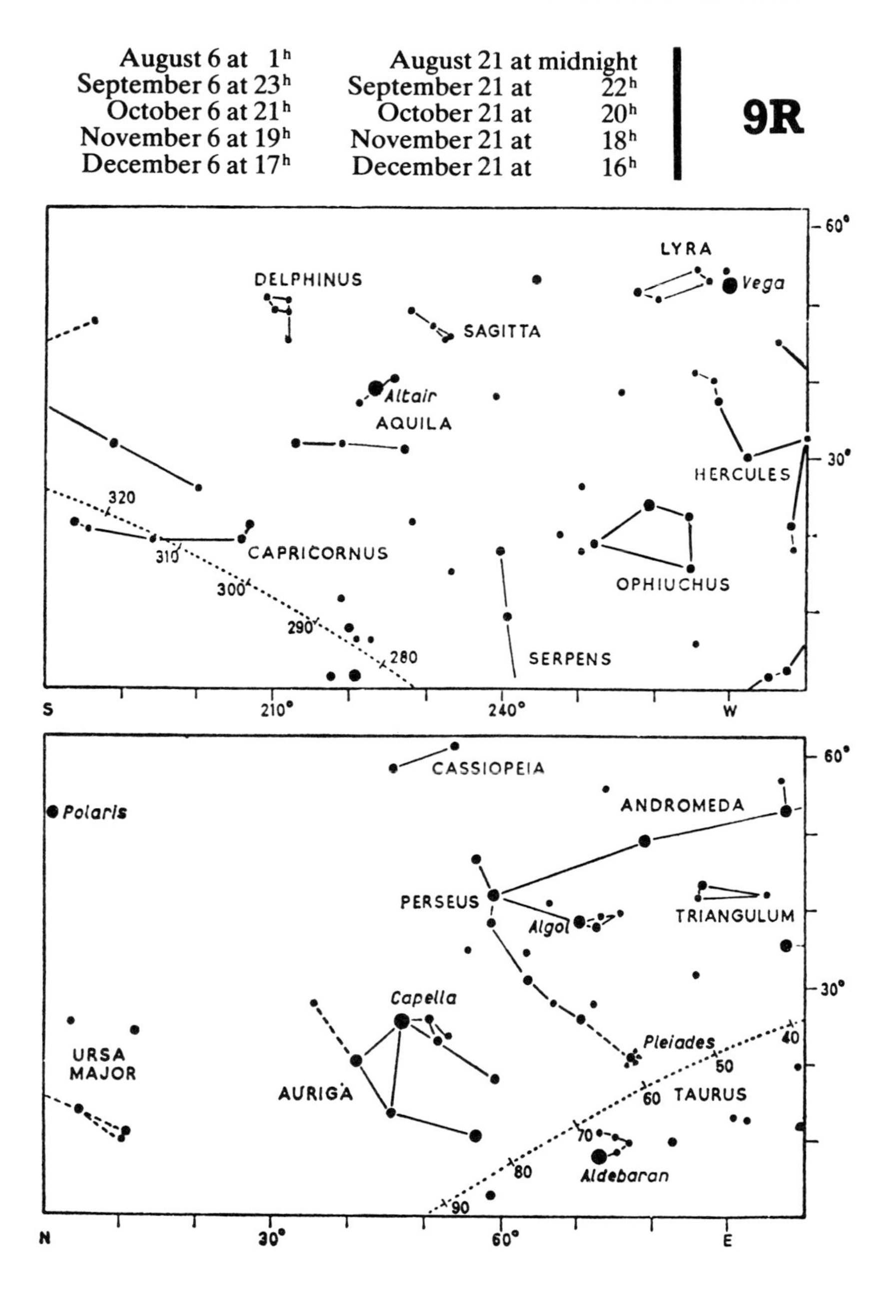
August 6 at 1h
September 6 at 23h
October 6 at 21h
November 6 at 19h
December 6 at 17h
August 21 at midnight
September 21 at 22h
October 21 at 20h
November 21 at 18h
December 21 at 16h
9R
LYRA
Vega
DELPHINUS
SAGITTA
Altair
AQUILA
HERCULES
CAPRICORNUS
OPHIUCHUS
SERPENS
320
310
300
290
280
60°
30°
S
210°
240°
W
CASSIOPEIA
Polaris
ANDROMEDA
PERSEUS
Algol
TRIANGULUM
Capella
Pleiades
URSA
MAJOR
AURIGA
TAURUS
Aldebaran
40
50
60
70
80
90
N
30°
60°
E

10L

August 6 at 3h	August 21 at 2h
September 6 at 1h	September 21 at midnight
October 6 at 23h	October 21 at 22h
November 6 at 21h	November 21 at 20h
December 6 at 19h	December 21 at 18h

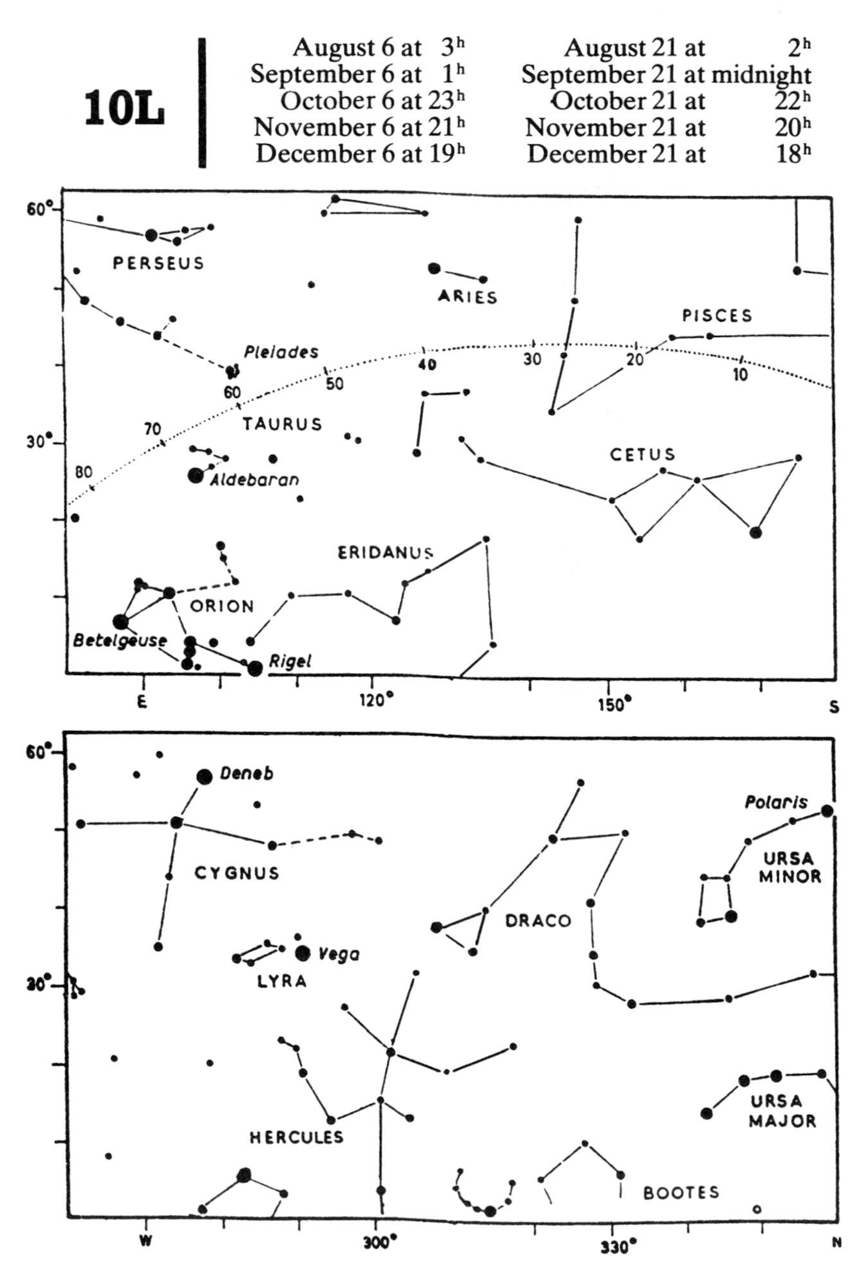

August 6 at 3^h	August 21 at 2^h	**10R**
September 6 at 1^h	September 21 at midnight	
October 6 at 23^h	October 21 at 22^h	
November 6 at 21^h	November 21 at 20^h	
December 6 at 19^h	December 21 at 18^h	

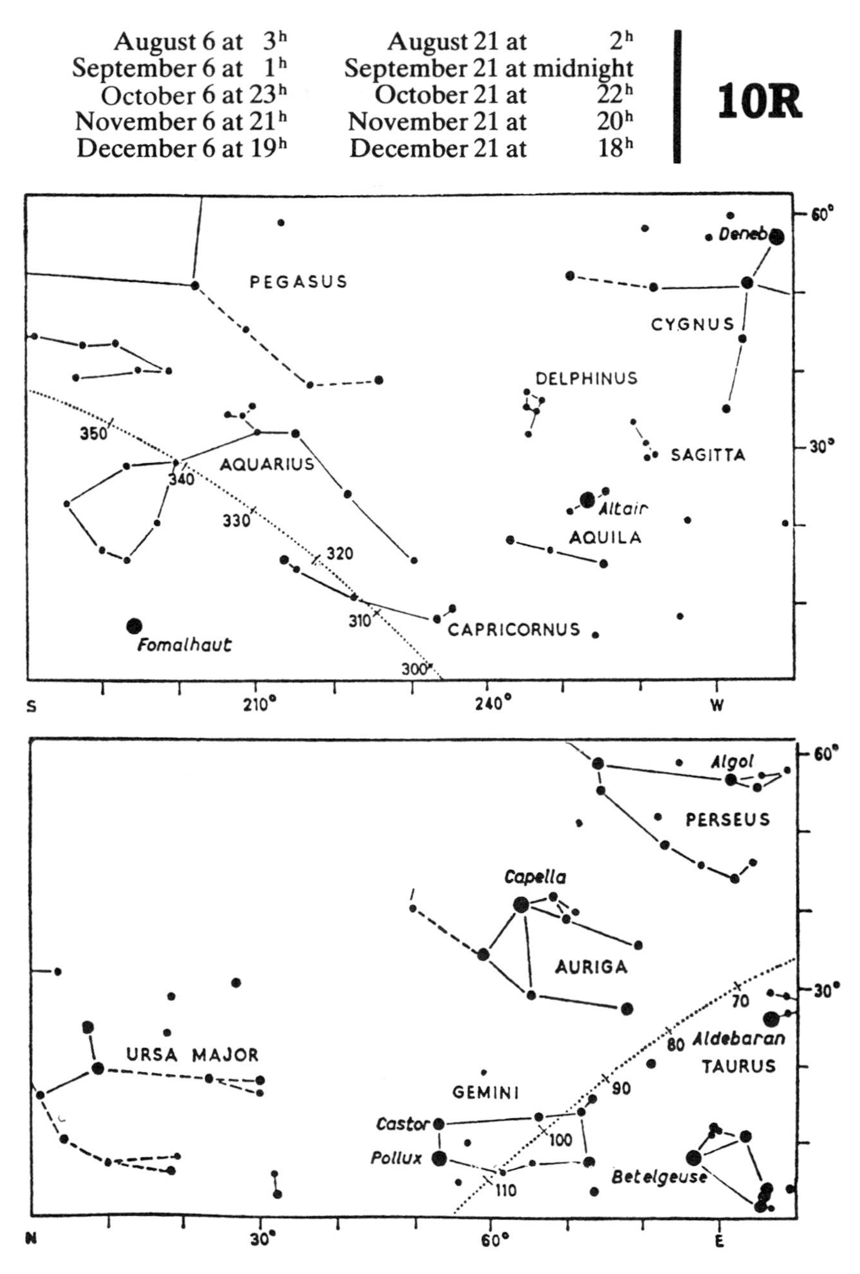

11L

September 6 at 3h	September 21 at 2h
October 6 at 1h	October 21 at midnight
November 6 at 23h	November 21 at 22h
December 6 at 21h	December 21 at 20h
January 6 at 19h	January 21 at 18h

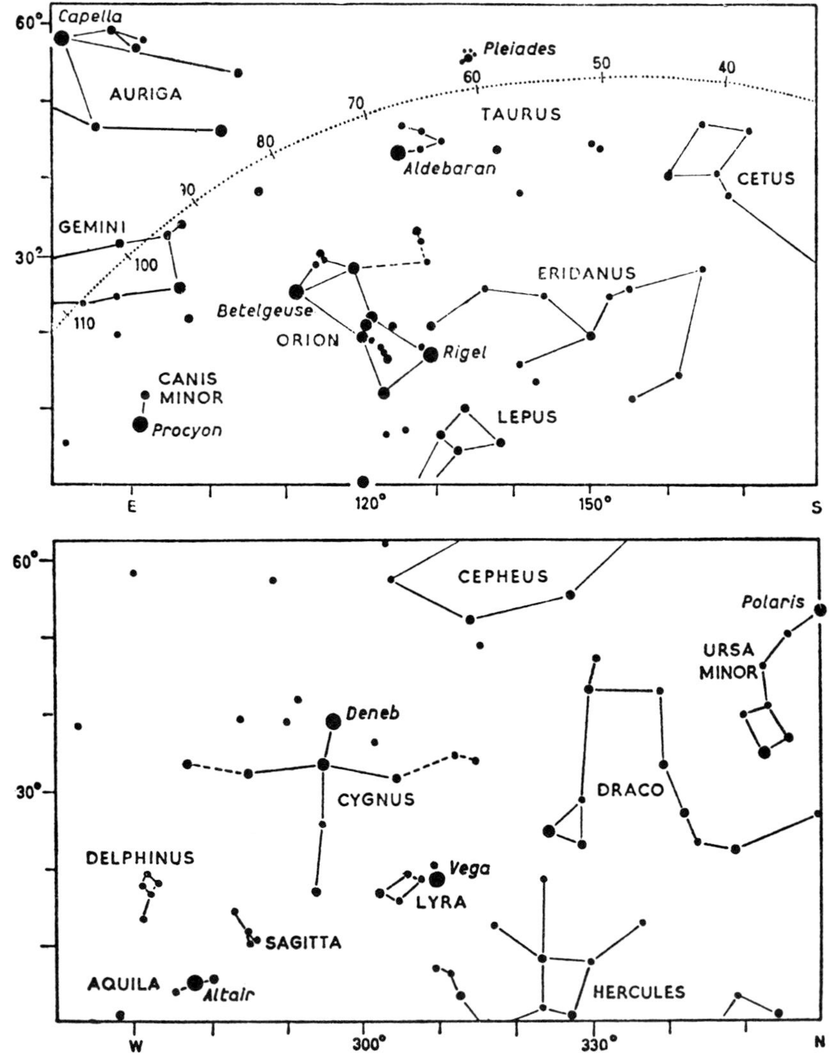

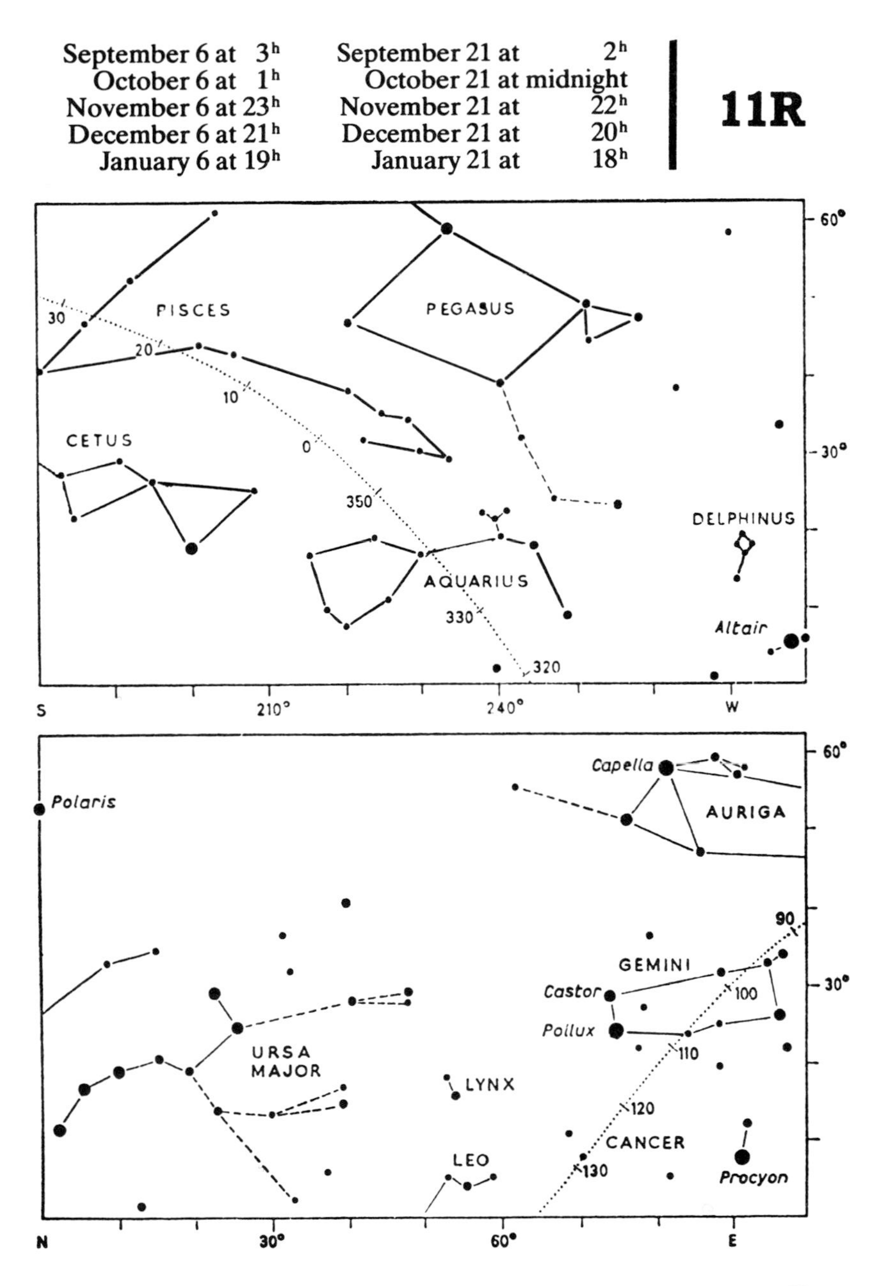
September 6 at 3h
October 6 at 1h
November 6 at 23h
December 6 at 21h
January 6 at 19h
September 21 at 2h
October 21 at midnight
November 21 at 22h
December 21 at 20h
January 21 at 18h
11R
PISCES
PEGASUS
CETUS
DELPHINUS
AQUARIUS
Altair
30
20
10
0
350
330
320
60°
30°
S
210°
240°
W
Capella
AURIGA
Polaris
GEMINI
Castor
Pollux
URSA MAJOR
LYNX
CANCER
LEO
Procyon
90
100
110
120
130
N
30°
60°
E

12L

October 6 at 3h	October 21 at 2h
November 6 at 1h	November 21 at midnight
December 6 at 23h	December 21 at 22h
January 6 at 21h	January 21 at 20h
February 6 at 19h	February 21 at 18h

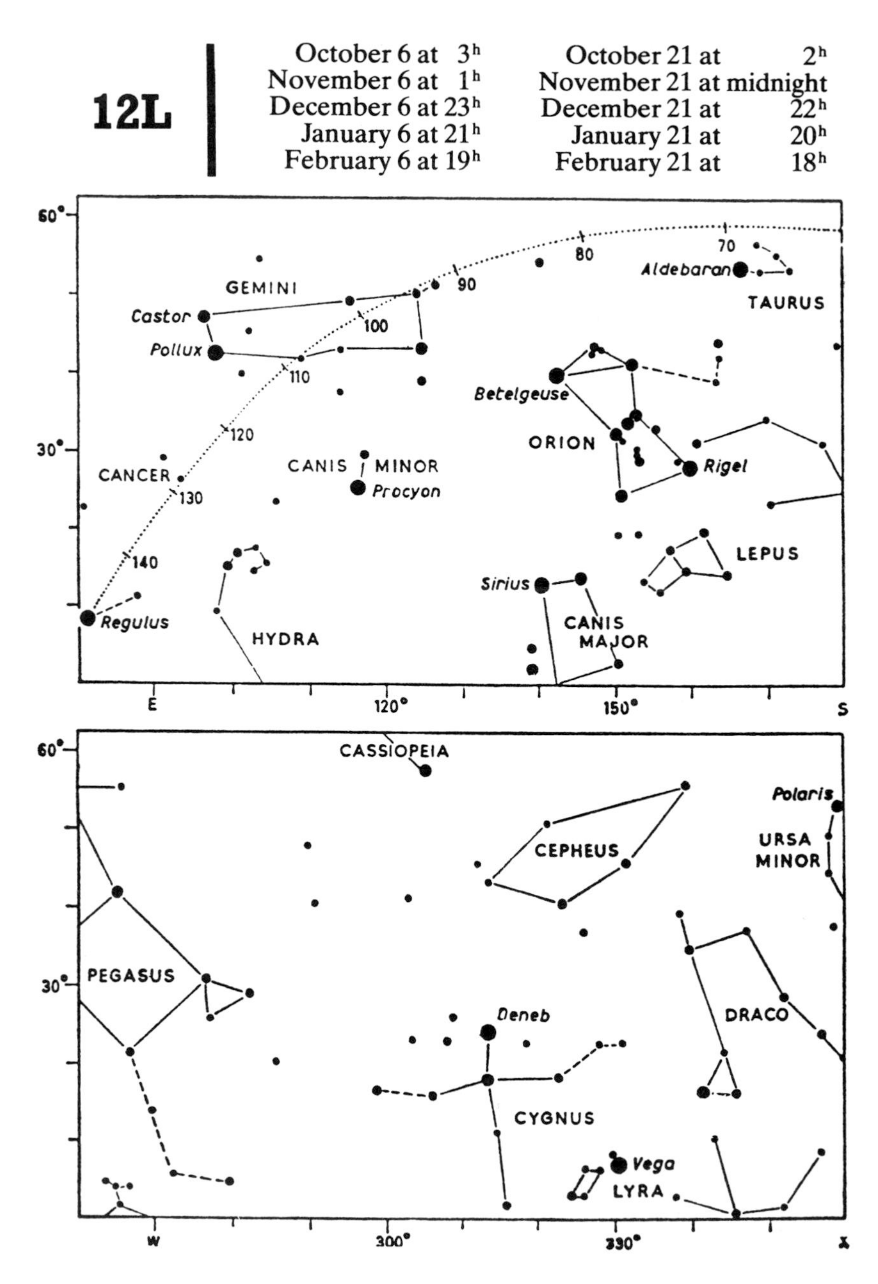

October 6 at 3h	October 21 at 2h	**12R**
November 6 at 1h	November 21 at midnight	
December 6 at 23h	December 21 at 22h	
January 6 at 21h	January 21 at 20h	
February 6 at 19h	February 21 at 18h	

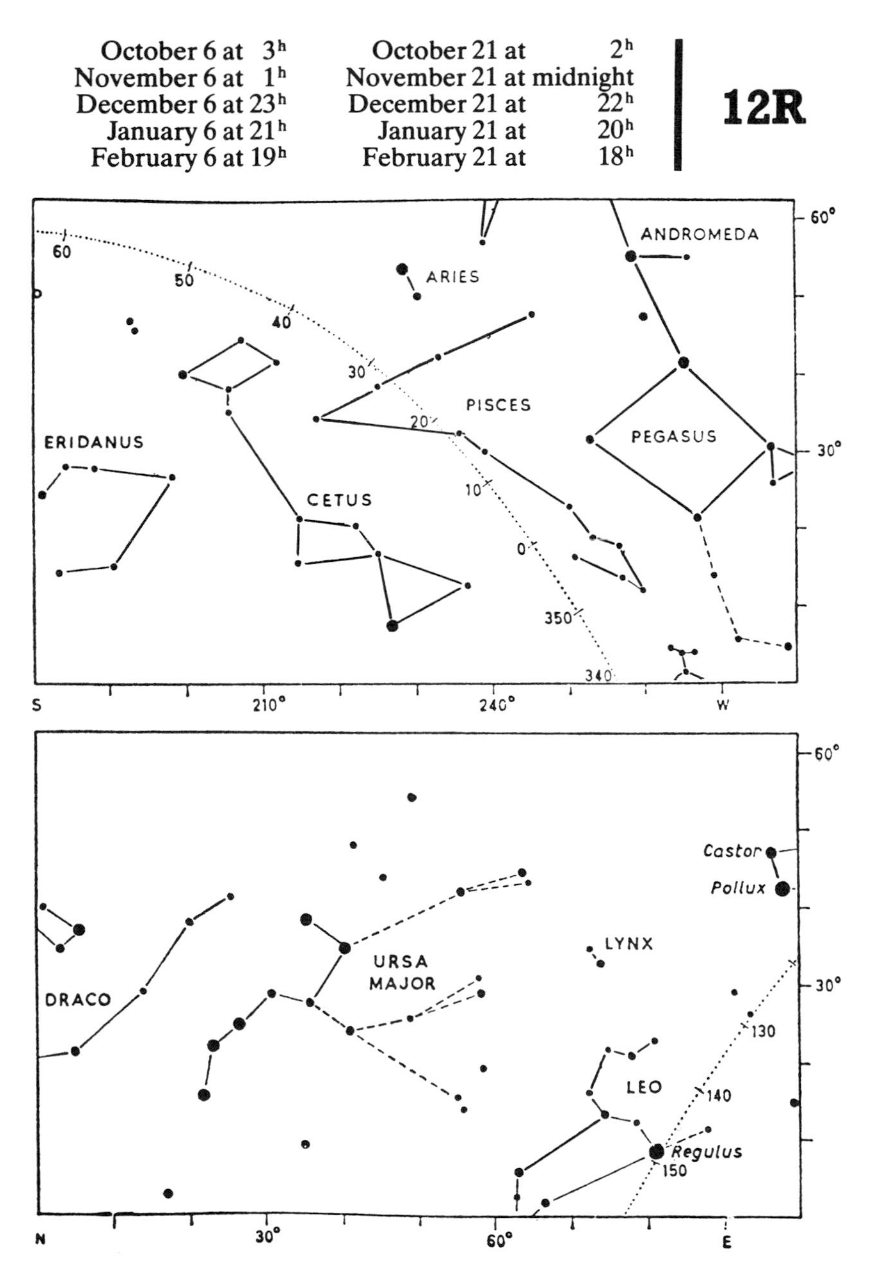

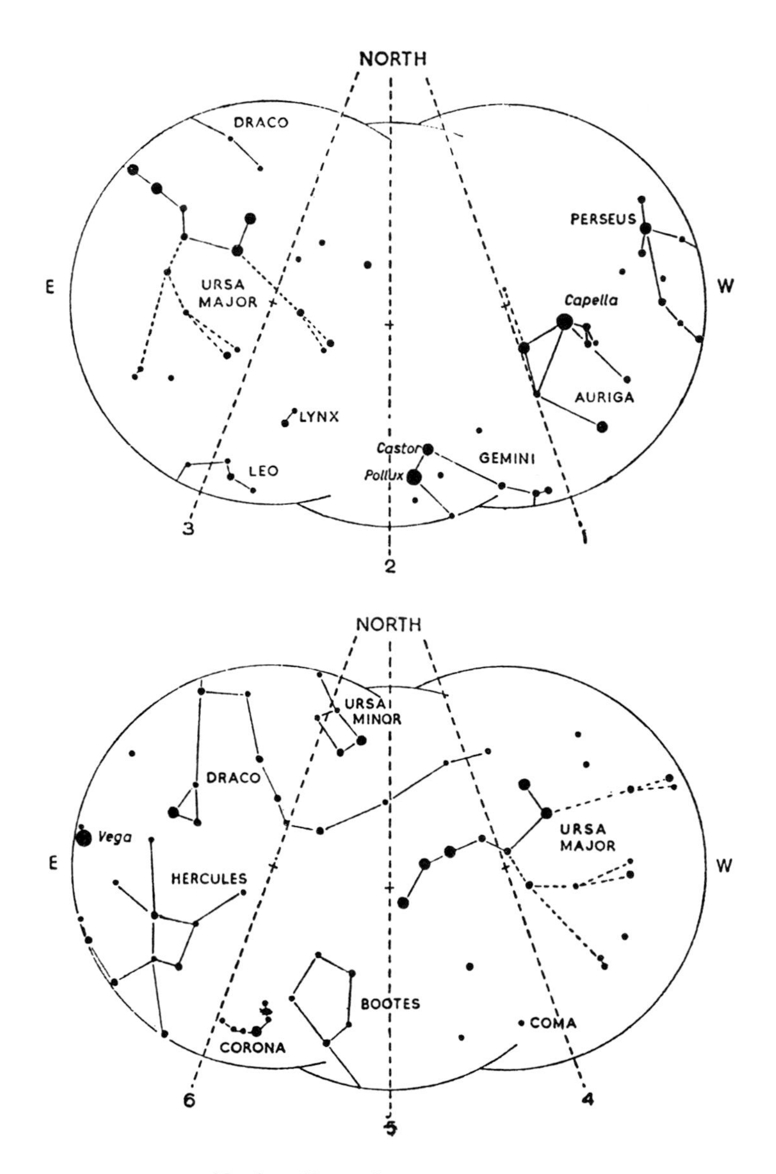

Northern Hemisphere Overhead Stars

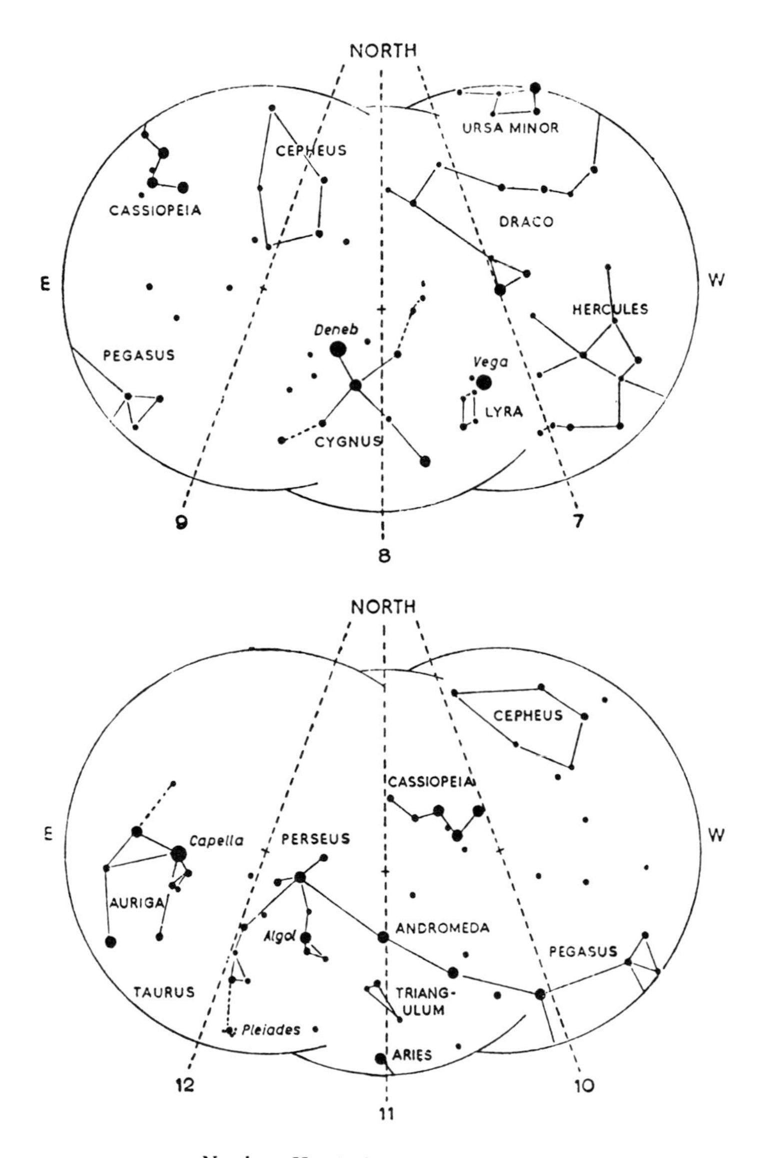

Northern Hemisphere Overhead Stars

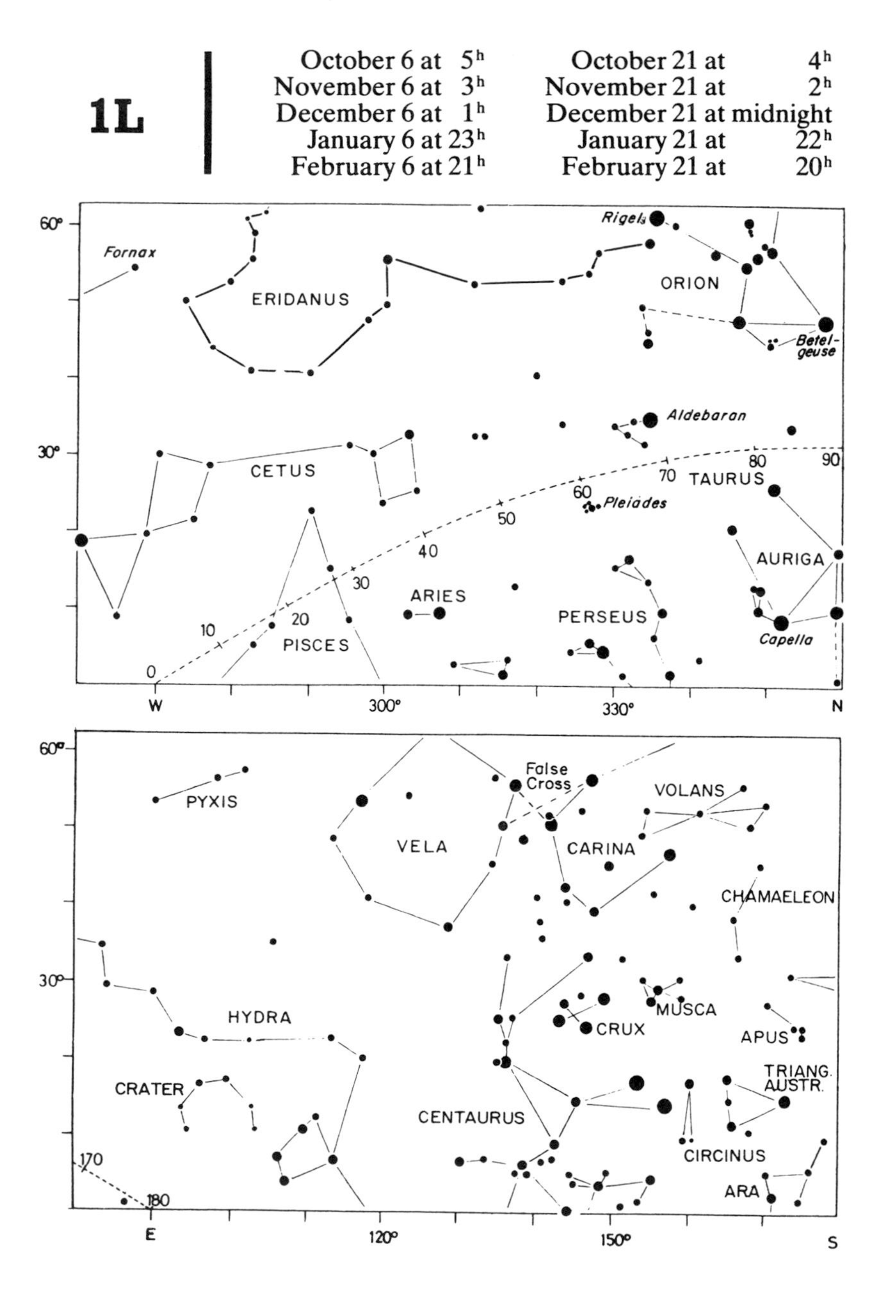
1L
October 6 at 5h
November 6 at 3h
December 6 at 1h
January 6 at 23h
February 6 at 21h
October 21 at 4h
November 21 at 2h
December 21 at midnight
January 21 at 22h
February 21 at 20h
60°
30°
Fornax
ERIDANUS
Rigel
ORION
Betelgeuse
Aldebaran
CETUS
TAURUS
Pleiades
AURIGA
ARIES
PISCES
PERSEUS
Capella
0
10
20
30
40
50
60
70
80
90
W
300°
330°
N
PYXIS
False Cross
VOLANS
VELA
CARINA
CHAMAELEON
HYDRA
MUSCA
CRUX
APUS
CRATER
TRIANG. AUSTR.
CENTAURUS
CIRCINUS
ARA
170
180
E
120°
150°
S

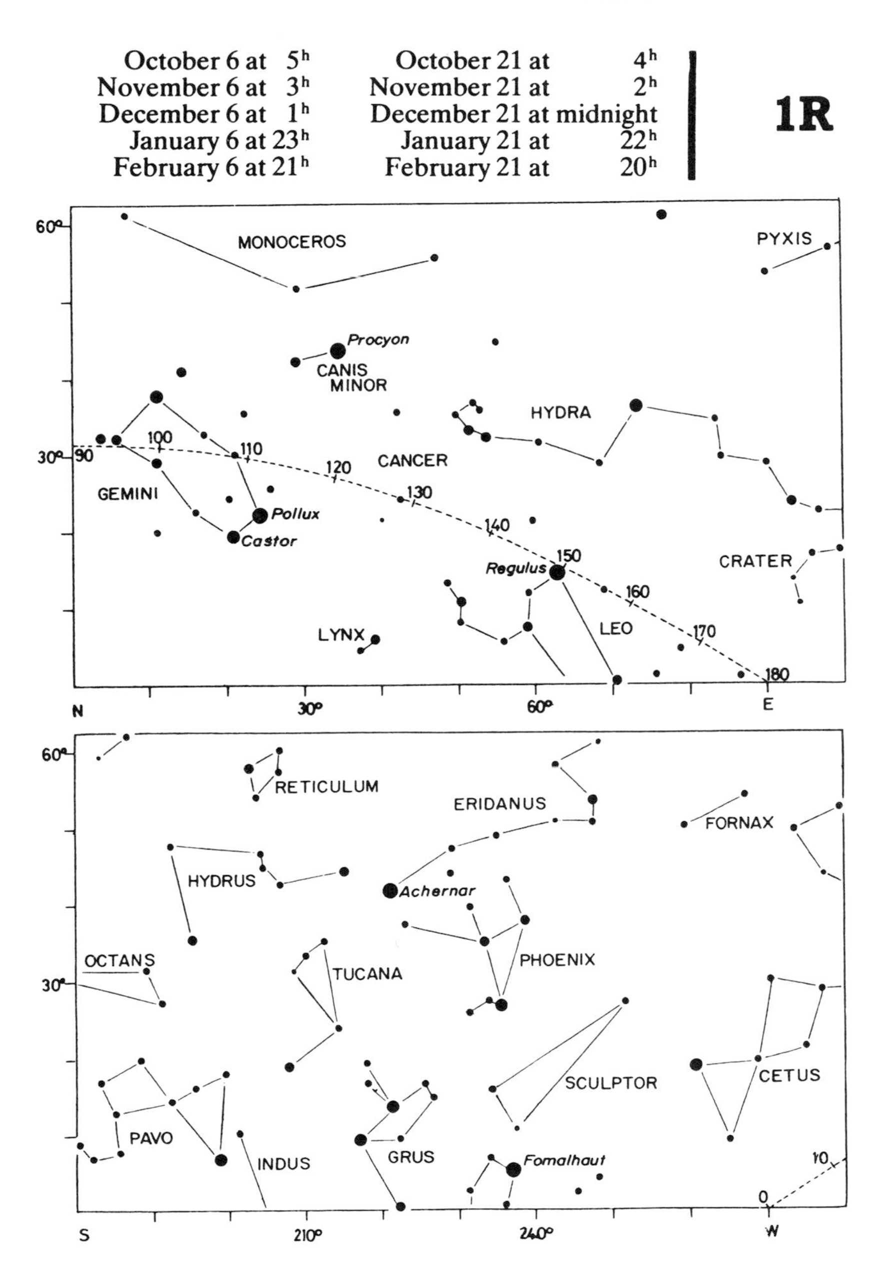
October 6 at 5h
November 6 at 3h
December 6 at 1h
January 6 at 23h
February 6 at 21h
October 21 at 4h
November 21 at 2h
December 21 at midnight
January 21 at 22h
February 21 at 20h
1R
MONOCEROS
PYXIS
Procyon
CANIS MINOR
HYDRA
CANCER
GEMINI
Pollux
Castor
Regulus
CRATER
LEO
LYNX
60°
30°
90
100
110
120
130
140
150
160
170
180
N
30°
60°
E
RETICULUM
ERIDANUS
FORNAX
HYDRUS
Achernar
PHOENIX
OCTANS
TUCANA
SCULPTOR
CETUS
PAVO
INDUS
GRUS
Fomalhaut
10
0
S
210°
240°
W

2L

November 6 at	5h	November 21 at	4h
December 6 at	3h	December 21 at	2h
January 6 at	1h	January 21 at	midnight
February 6 at	23h	February 21 at	22h
March 6 at	21h	March 21 at	20h

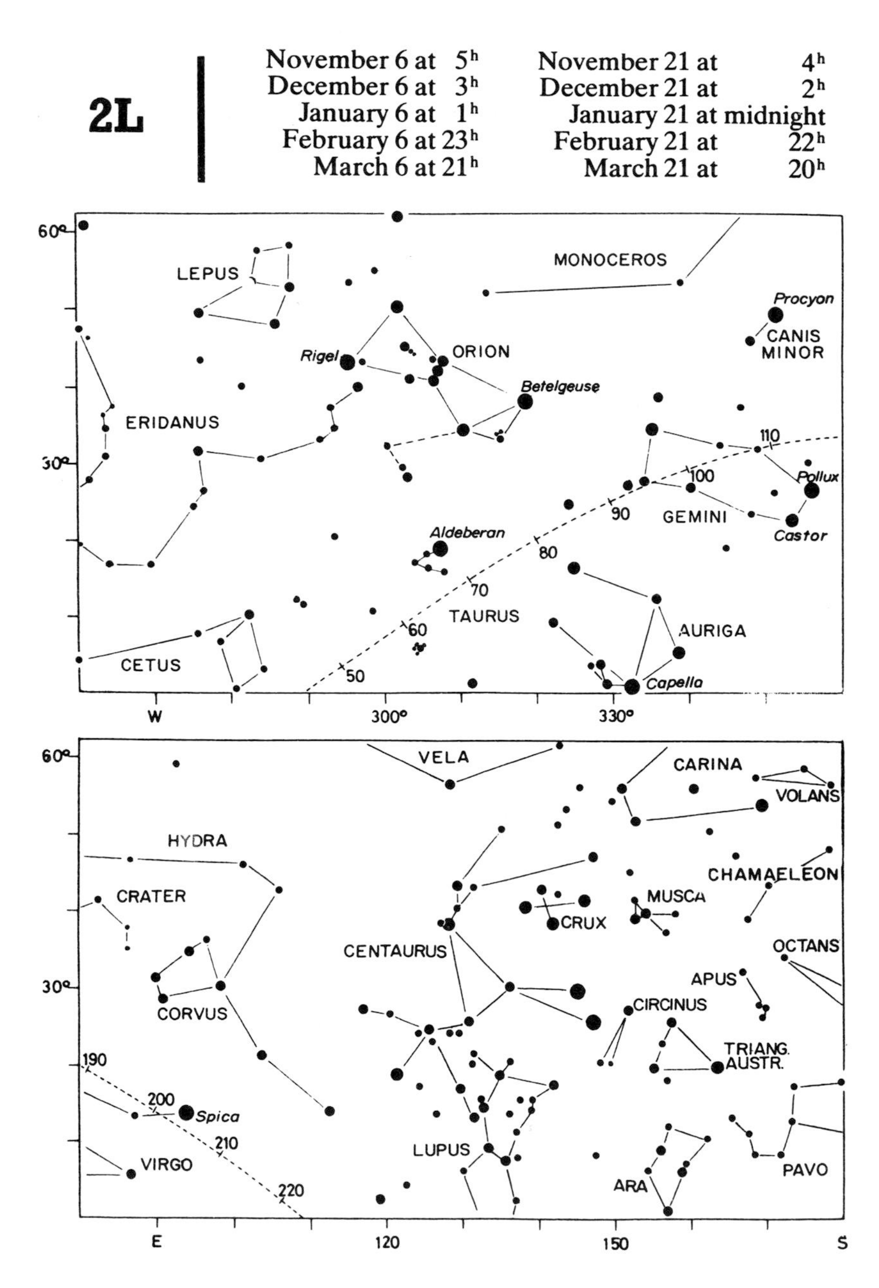

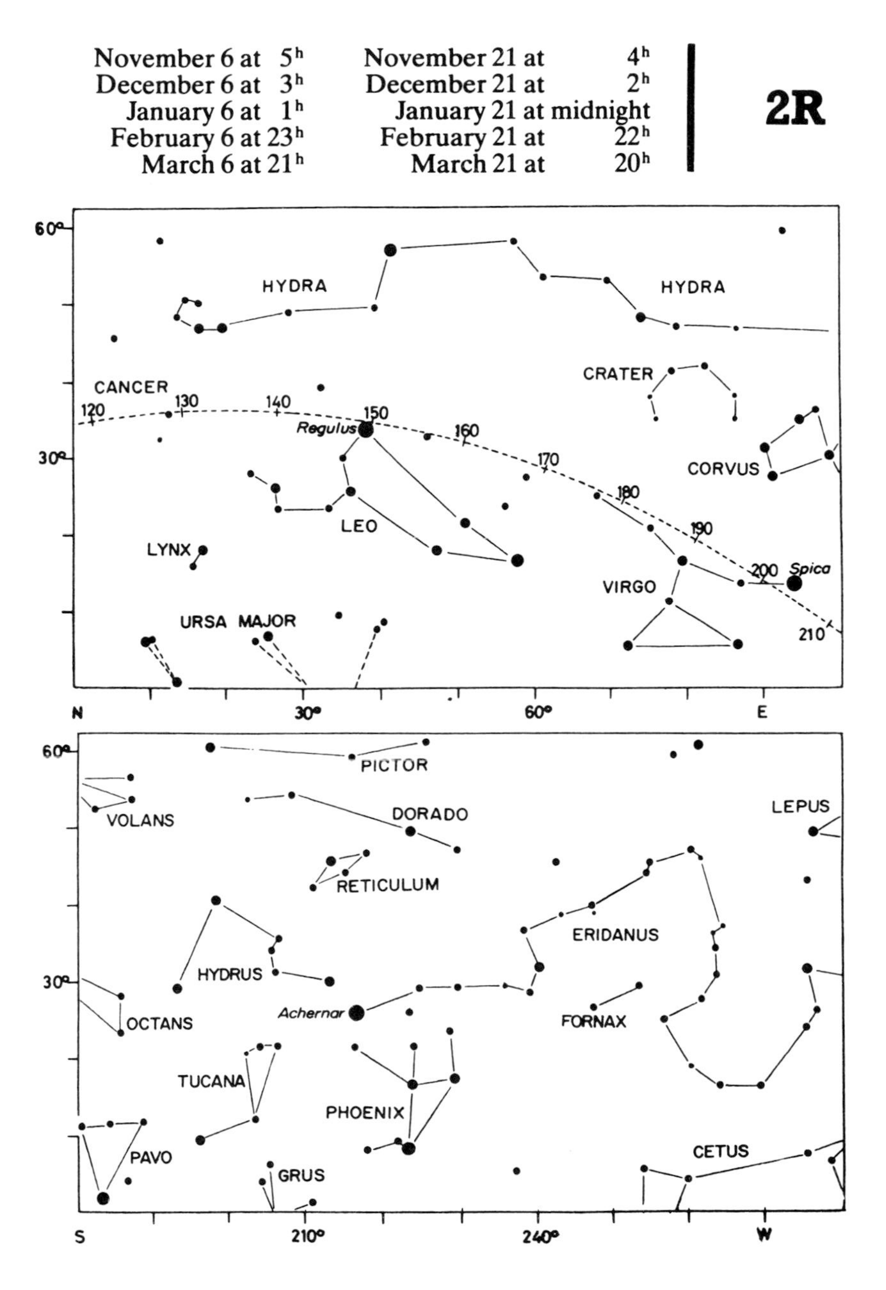
November 6 at 5h
December 6 at 3h
January 6 at 1h
February 6 at 23h
March 6 at 21h
November 21 at 4h
December 21 at 2h
January 21 at midnight
February 21 at 22h
March 21 at 20h
2R
60°
30°
HYDRA
HYDRA
CRATER
CANCER
120
130
140
150
160
170
180
190
200
210
Regulus
Spica
CORVUS
LEO
LYNX
VIRGO
URSA MAJOR
N
30°
60°
E
60°
PICTOR
VOLANS
DORADO
LEPUS
RETICULUM
ERIDANUS
30°
HYDRUS
OCTANS
Achernar
FORNAX
TUCANA
PHOENIX
CETUS
PAVO
GRUS
S
210°
240°
W

3L

January 6 at 3^h	January 21 at 2^h
February 6 at 1^h	February 21 at midnight
March 6 at 23^h	March 21 at 22^h
April 6 at 21^h	April 21 at 20^h
May 6 at 19^h	May 21 at 18^h

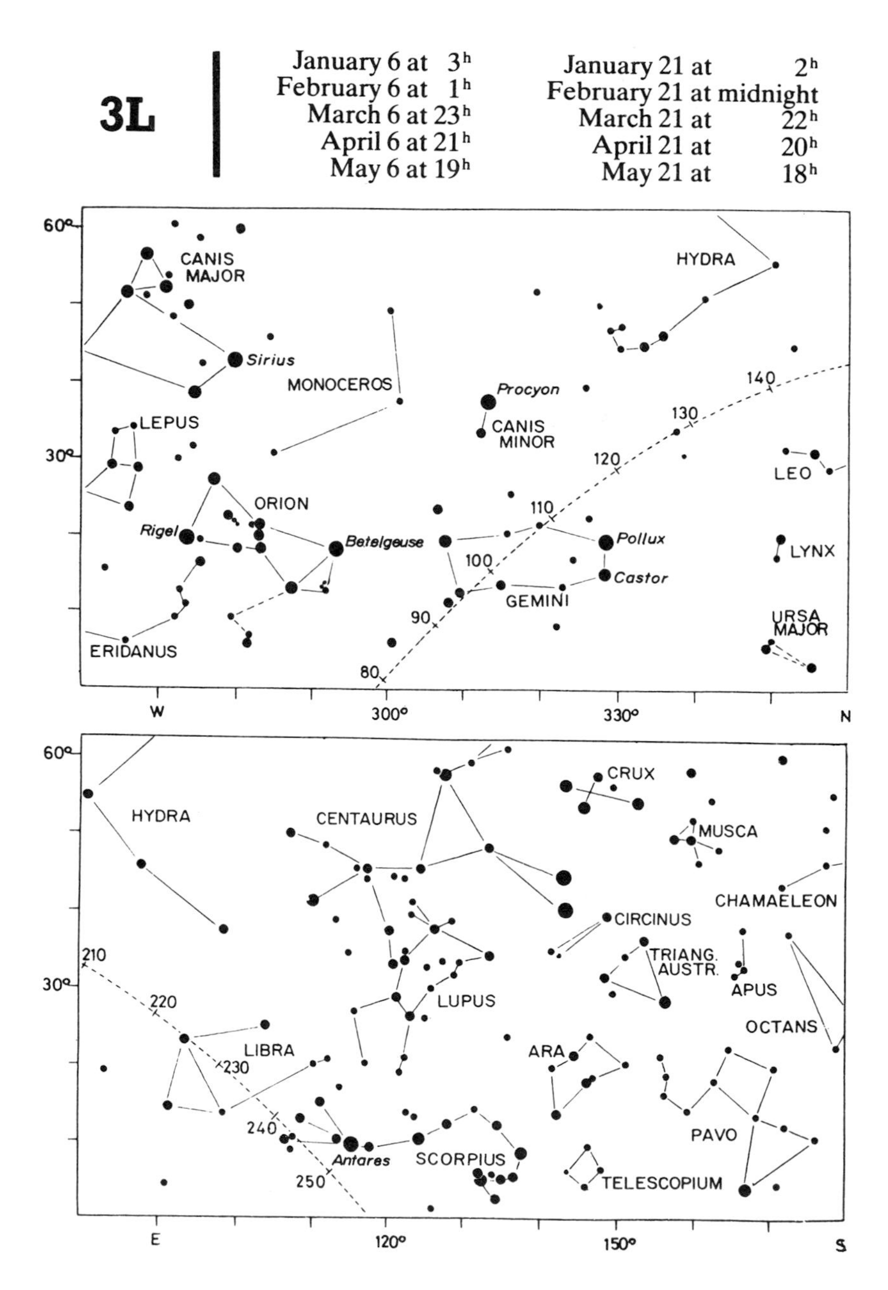

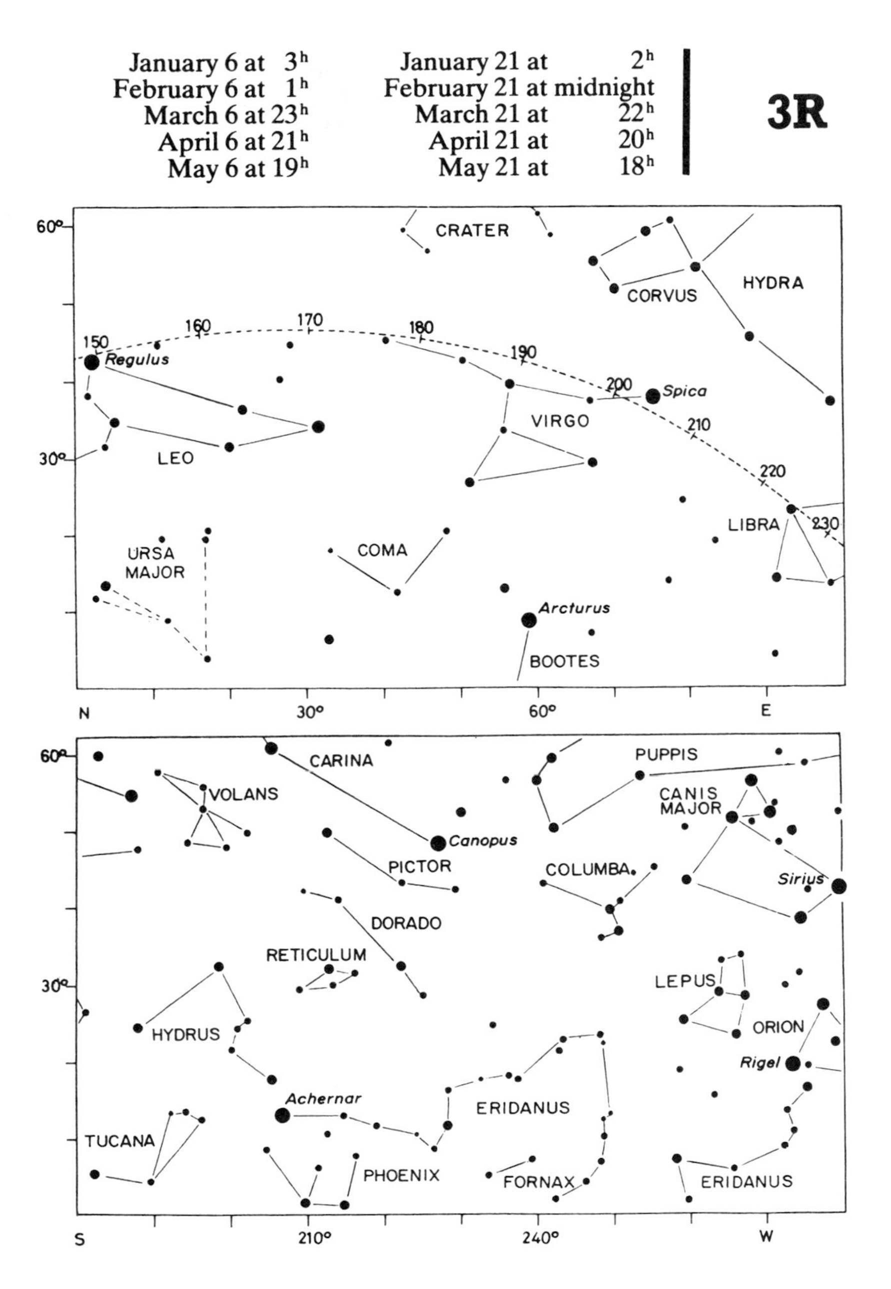
January 6 at 3h
February 6 at 1h
March 6 at 23h
April 6 at 21h
May 6 at 19h
January 21 at 2h
February 21 at midnight
March 21 at 22h
April 21 at 20h
May 21 at 18h
3R
60°
30°
N
30°
60°
E
150
160
170
180
190
200
210
220
230
CRATER
CORVUS
HYDRA
Regulus
Spica
VIRGO
LEO
LIBRA
URSA MAJOR
COMA
Arcturus
BOOTES
S
210°
240°
W
CARINA
PUPPIS
VOLANS
CANIS MAJOR
Canopus
PICTOR
COLUMBA
Sirius
DORADO
RETICULUM
LEPUS
ORION
HYDRUS
Rigel
Achernar
ERIDANUS
TUCANA
PHOENIX
FORNAX
ERIDANUS

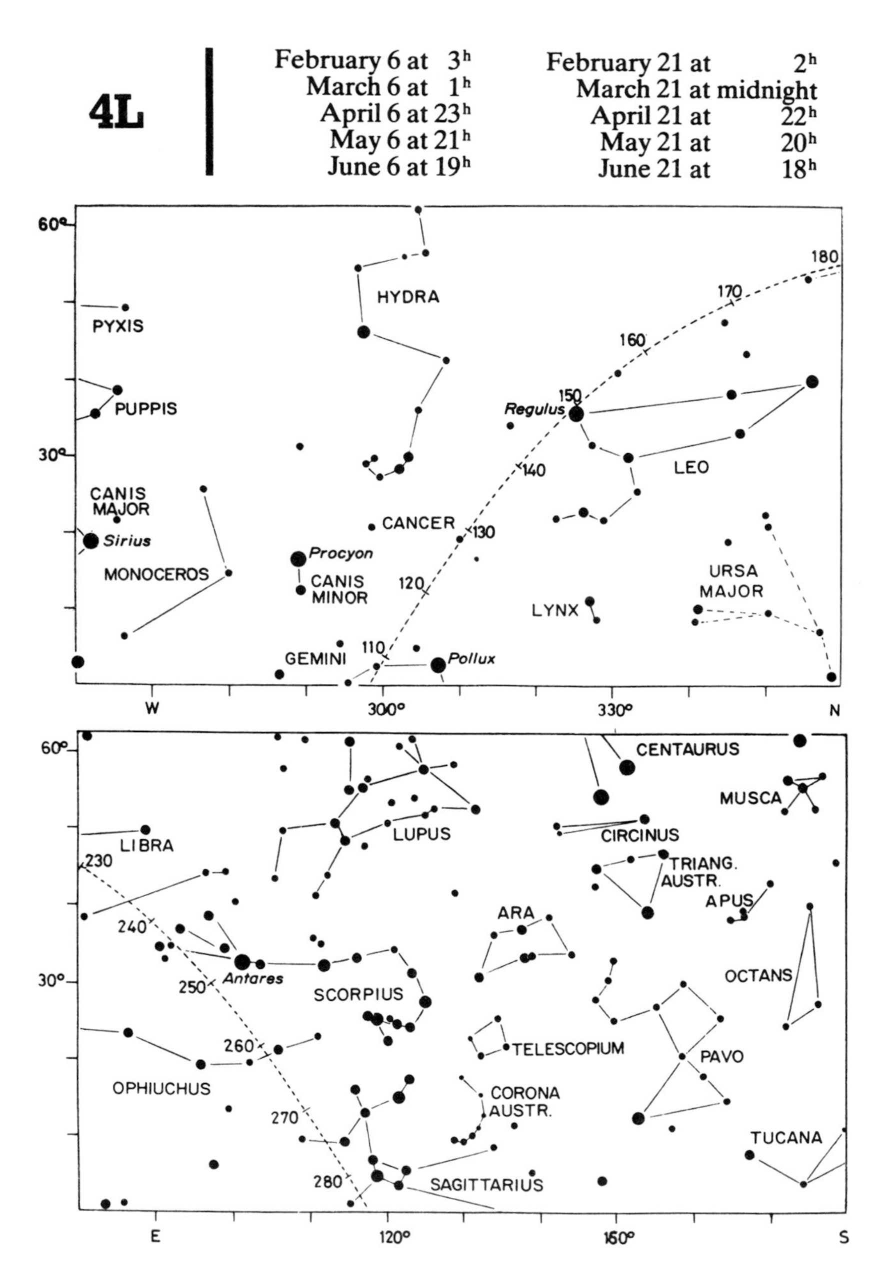
4L
February 6 at 3h
March 6 at 1h
April 6 at 23h
May 6 at 21h
June 6 at 19h
February 21 at 2h
March 21 at midnight
April 21 at 22h
May 21 at 20h
June 21 at 18h
60°
30°
HYDRA
PYXIS
PUPPIS
CANIS MAJOR
Sirius
MONOCEROS
Procyon
CANIS MINOR
CANCER
GEMINI
Pollux
Regulus
LEO
LYNX
URSA MAJOR
110
120
130
140
150
160
170
180
W
300°
330°
N
CENTAURUS
MUSCA
LIBRA
LUPUS
CIRCINUS
TRIANG. AUSTR.
APUS
ARA
OCTANS
Antares
SCORPIUS
TELESCOPIUM
PAVO
OPHIUCHUS
CORONA AUSTR.
TUCANA
SAGITTARIUS
230
240
250
260
270
280
E
120°
160°
S

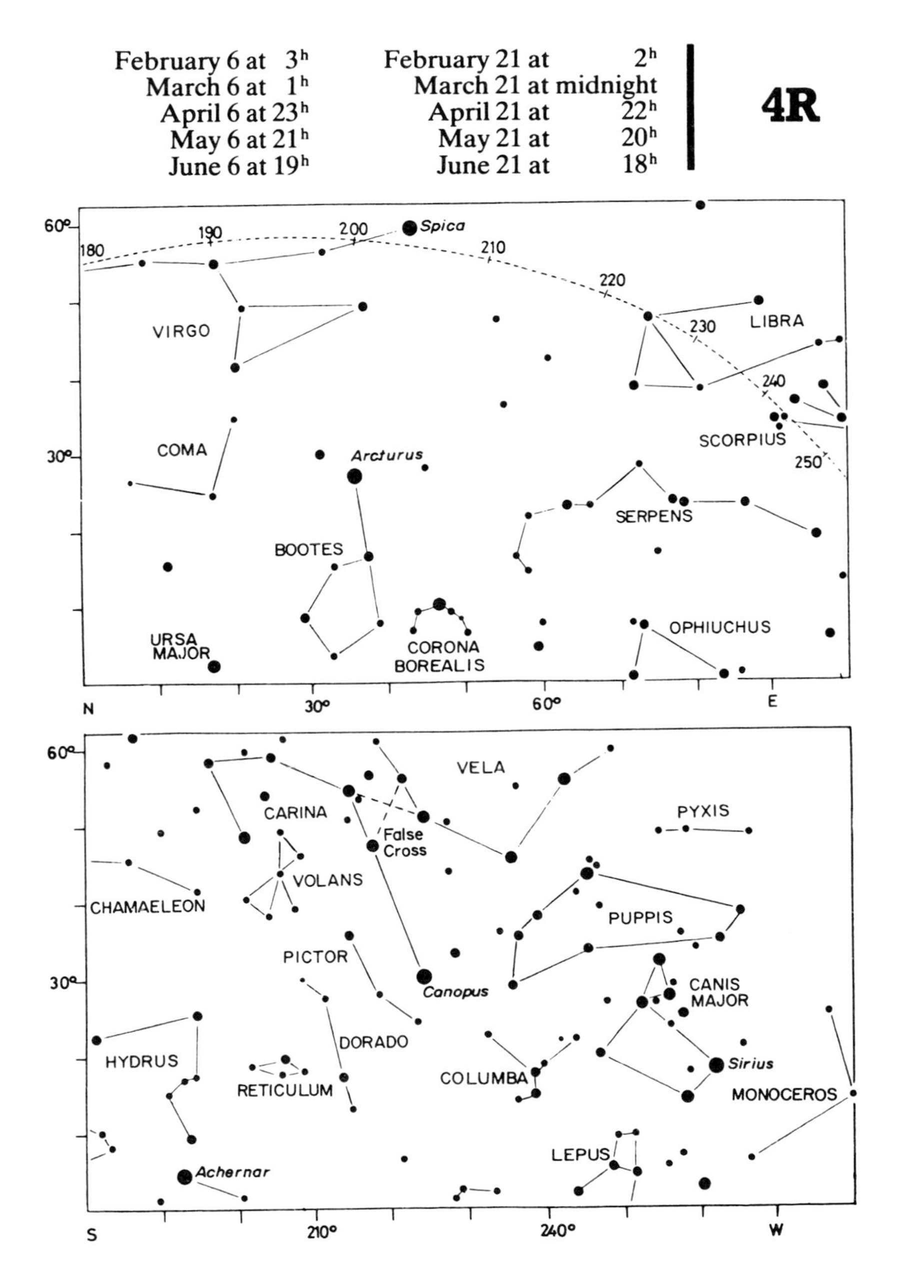
February 6 at 3^h
March 6 at 1^h
April 6 at 23^h
May 6 at 21^h
June 6 at 19^h
February 21 at 2^h
March 21 at midnight
April 21 at 22^h
May 21 at 20^h
June 21 at 18^h
4R
60°
30°
180
190
200
210
220
230
240
250
Spica
VIRGO
LIBRA
COMA
Arcturus
SCORPIUS
SERPENS
BOOTES
URSA MAJOR
CORONA BOREALIS
OPHIUCHUS
N
30°
60°
E
60°
30°
VELA
CARINA
PYXIS
False Cross
VOLANS
CHAMAELEON
PUPPIS
PICTOR
Canopus
CANIS MAJOR
DORADO
HYDRUS
Sirius
RETICULUM
COLUMBA
MONOCEROS
LEPUS
Achernar
S
210°
240°
W

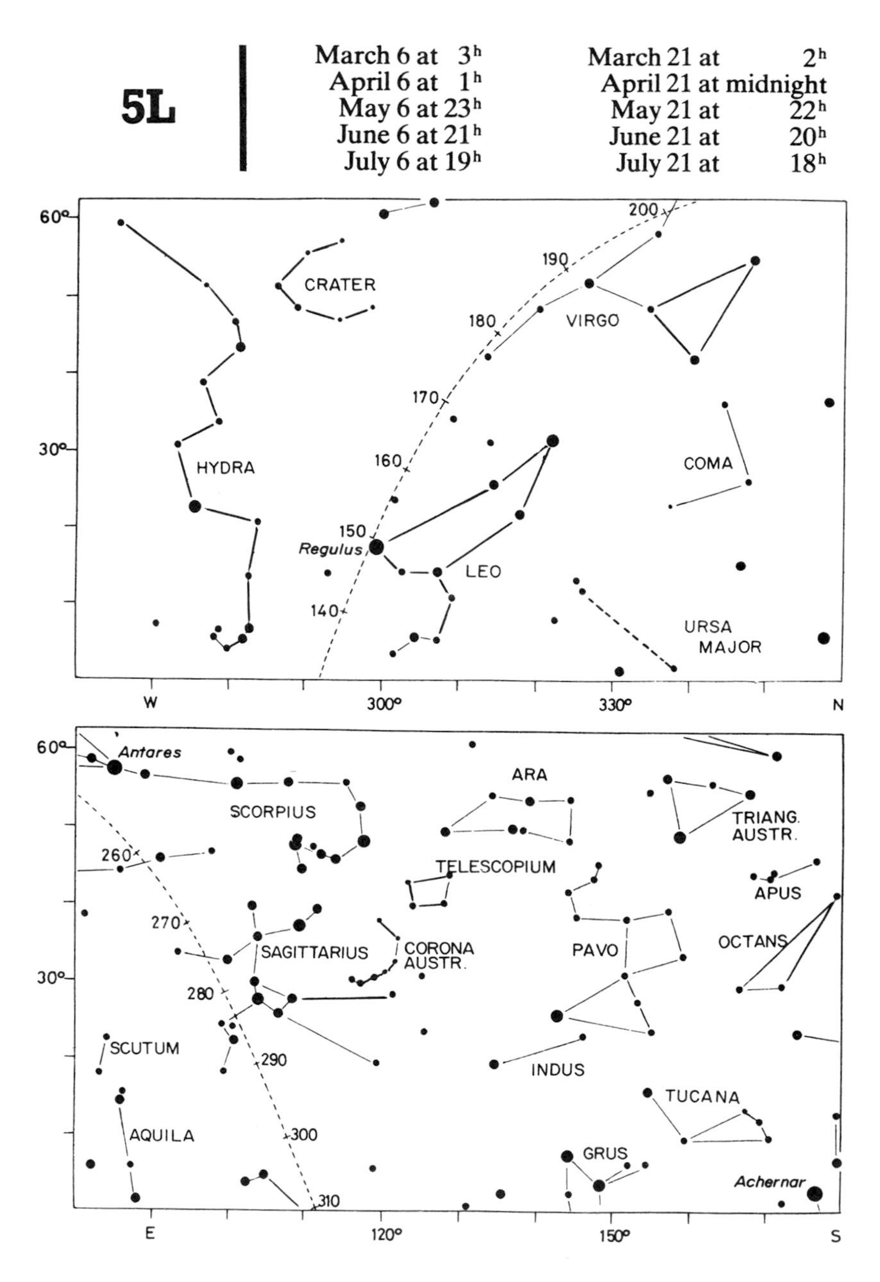

5L
March 6 at 3h
April 6 at 1h
May 6 at 23h
June 6 at 21h
July 6 at 19h
March 21 at 2h
April 21 at midnight
May 21 at 22h
June 21 at 20h
July 21 at 18h
60°
30°
CRATER
VIRGO
HYDRA
COMA
Regulus
LEO
URSA
MAJOR
200
190
180
170
160
150
140
W
300°
330°
N
Antares
ARA
SCORPIUS
TRIANG.
AUSTR.
TELESCOPIUM
APUS
OCTANS
SAGITTARIUS
CORONA
AUSTR.
PAVO
SCUTUM
INDUS
TUCANA
AQUILA
GRUS
Achernar
260
270
280
290
300
310
E
120°
150°
S

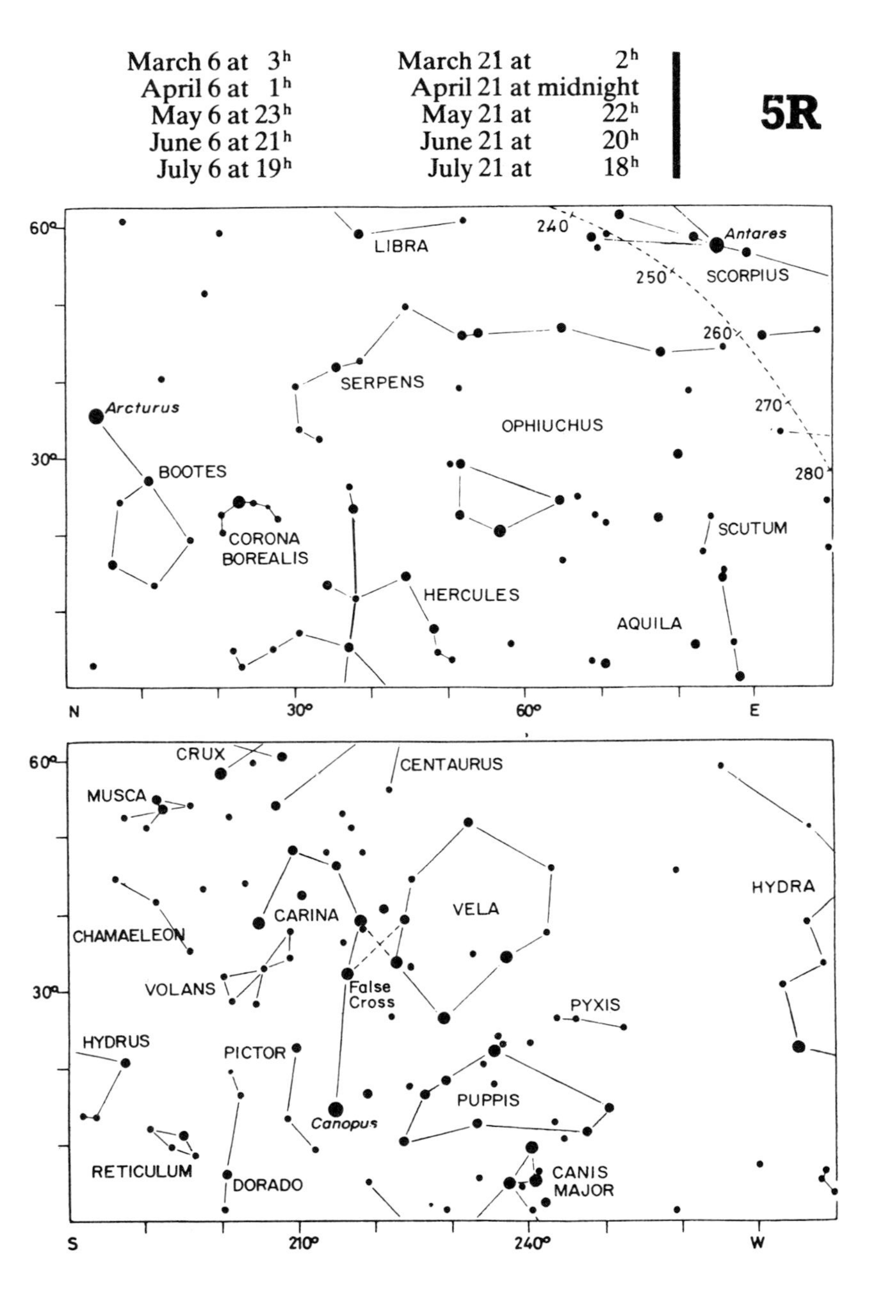
March 6 at 3h
April 6 at 1h
May 6 at 23h
June 6 at 21h
July 6 at 19h
March 21 at 2h
April 21 at midnight
May 21 at 22h
June 21 at 20h
July 21 at 18h
5R
60°
30°
LIBRA
Antares
SCORPIUS
240
250
260
270
280
SERPENS
OPHIUCHUS
Arcturus
BOOTES
CORONA
BOREALIS
HERCULES
SCUTUM
AQUILA
N
30°
60°
E
CRUX
CENTAURUS
MUSCA
HYDRA
CHAMAELEON
CARINA
VELA
VOLANS
False
Cross
PYXIS
HYDRUS
PICTOR
Canopus
PUPPIS
RETICULUM
DORADO
CANIS
MAJOR
S
210°
240°
W

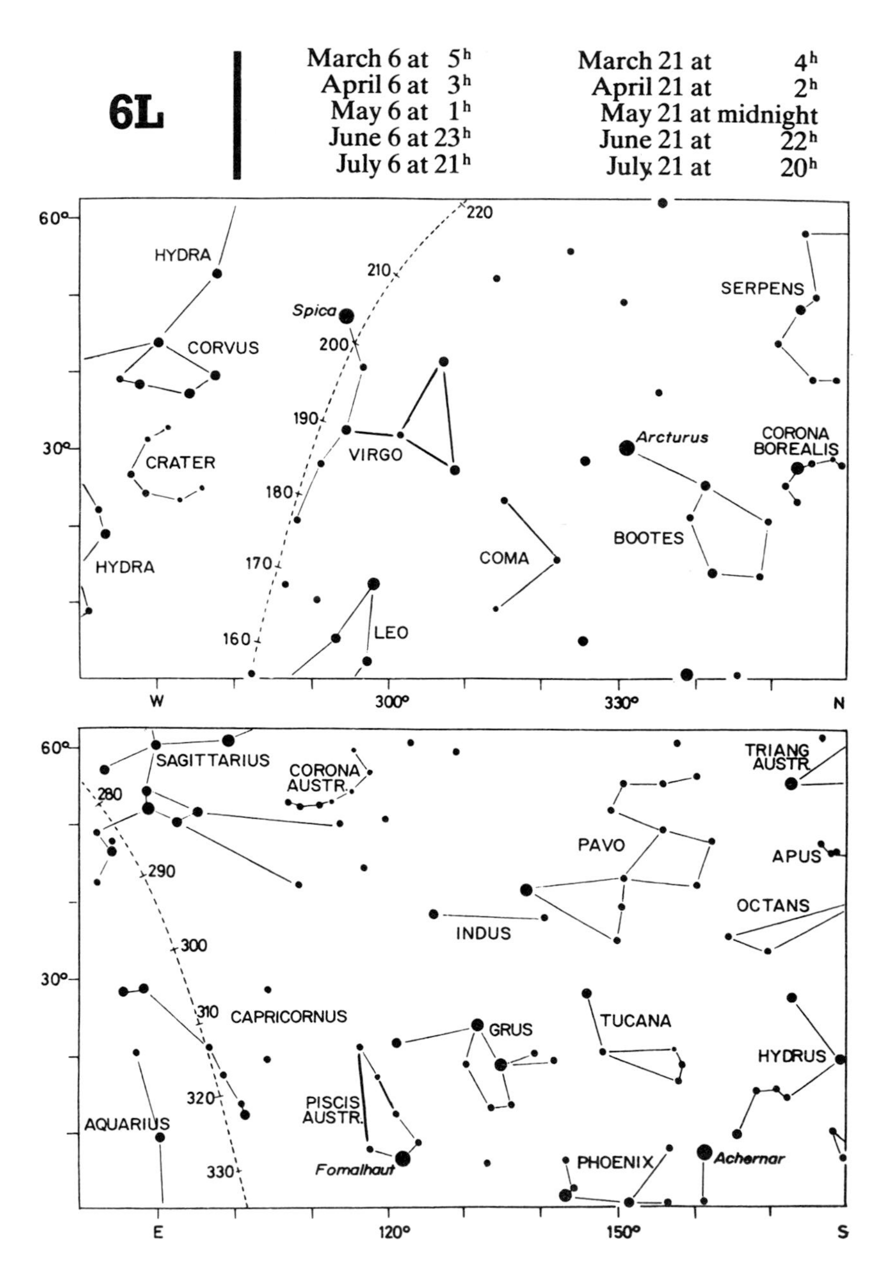
6L
March 6 at 5h
April 6 at 3h
May 6 at 1h
June 6 at 23h
July 6 at 21h
March 21 at 4h
April 21 at 2h
May 21 at midnight
June 21 at 22h
July 21 at 20h
60°
30°
HYDRA
CORVUS
CRATER
HYDRA
Spica
VIRGO
LEO
COMA
Arcturus
BOOTES
SERPENS
CORONA BOREALIS
220
210
200
190
180
170
160
W
300°
330°
N
SAGITTARIUS
CORONA AUSTR.
TRIANG AUSTR.
PAVO
APUS
OCTANS
INDUS
280
290
300
310
320
330
CAPRICORNUS
GRUS
TUCANA
HYDRUS
AQUARIUS
PISCIS AUSTR.
Fomalhaut
PHOENIX
Achernar
E
120°
150°
S

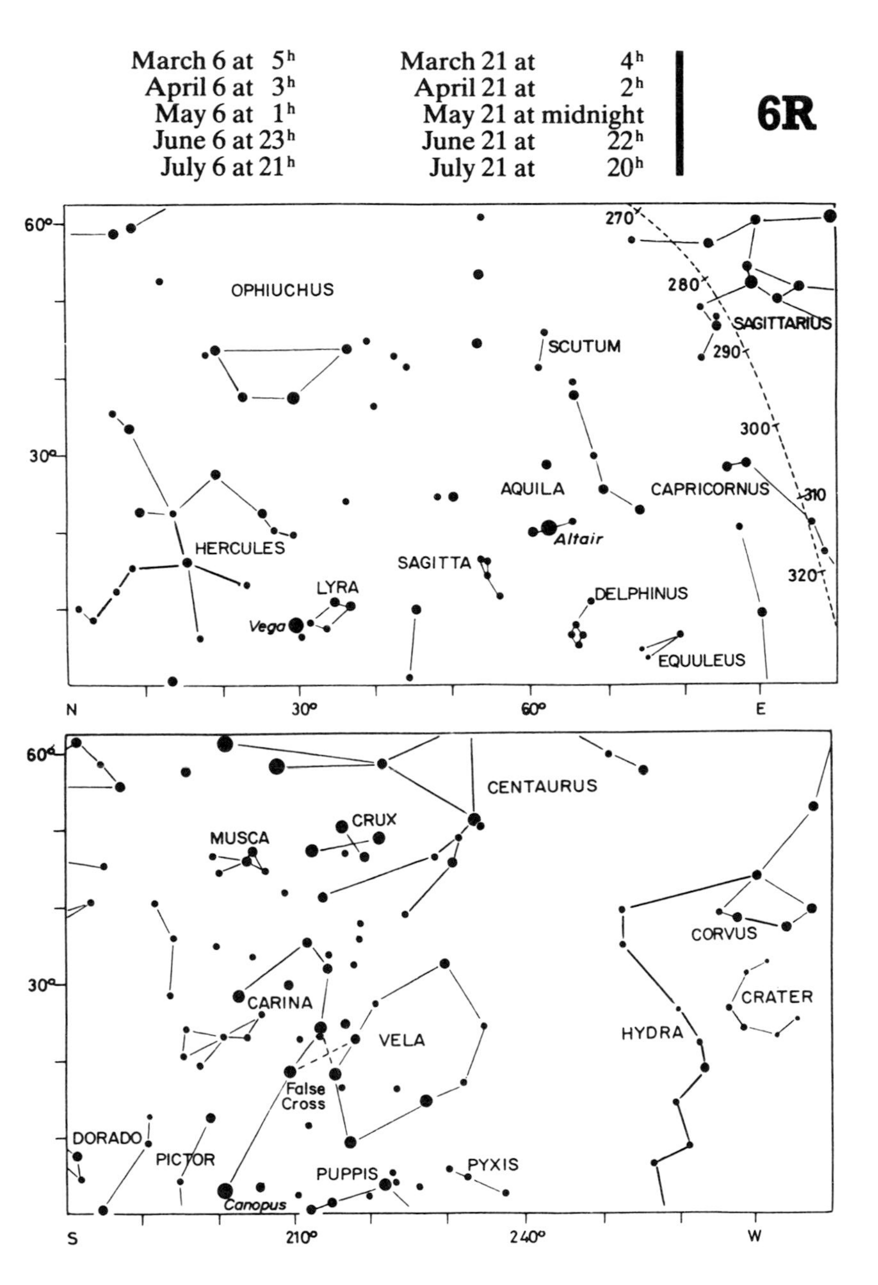

March 6 at 5h
April 6 at 3h
May 6 at 1h
June 6 at 23h
July 6 at 21h
March 21 at 4h
April 21 at 2h
May 21 at midnight
June 21 at 22h
July 21 at 20h
6R
60°
30°
270
280
290
300
310
320
OPHIUCHUS
SCUTUM
SAGITTARIUS
AQUILA
CAPRICORNUS
Altair
HERCULES
SAGITTA
LYRA
Vega
DELPHINUS
EQUULEUS
N
30°
60°
E
CENTAURUS
CRUX
MUSCA
CORVUS
CARINA
CRATER
VELA
HYDRA
False
Cross
DORADO
PICTOR
PUPPIS
PYXIS
Canopus
S
210°
240°
W

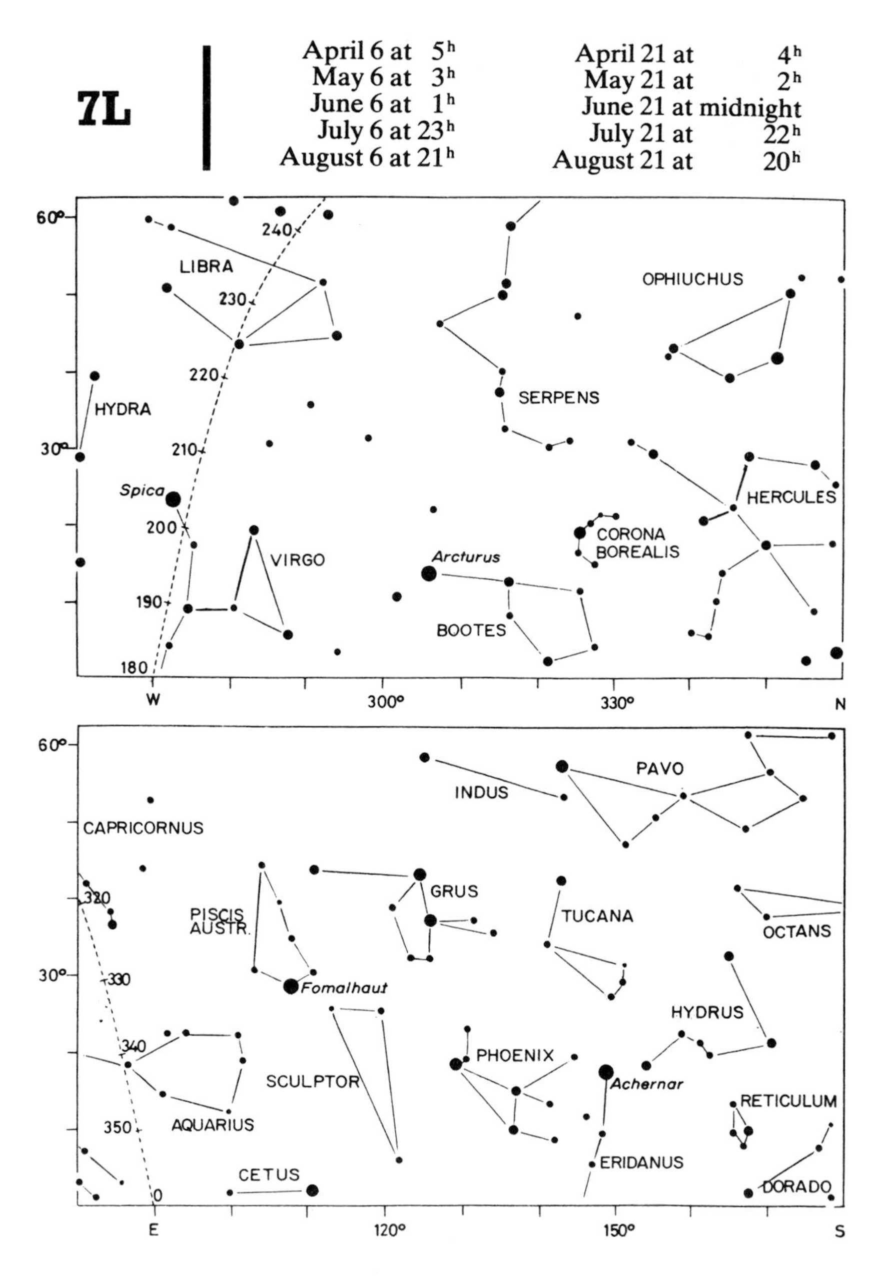
7L
April 6 at 5h
May 6 at 3h
June 6 at 1h
July 6 at 23h
August 6 at 21h
April 21 at 4h
May 21 at 2h
June 21 at midnight
July 21 at 22h
August 21 at 20h
60°
30°
240
230
220
210
200
190
180
LIBRA
HYDRA
Spica
VIRGO
OPHIUCHUS
SERPENS
HERCULES
CORONA BOREALIS
Arcturus
BOOTES
W
300°
330°
N
60°
30°
320
330
340
350
0
CAPRICORNUS
INDUS
PAVO
PISCIS AUSTR.
GRUS
TUCANA
OCTANS
Fomalhaut
HYDRUS
PHOENIX
SCULPTOR
Achernar
AQUARIUS
RETICULUM
ERIDANUS
CETUS
DORADO
E
120°
150°
S

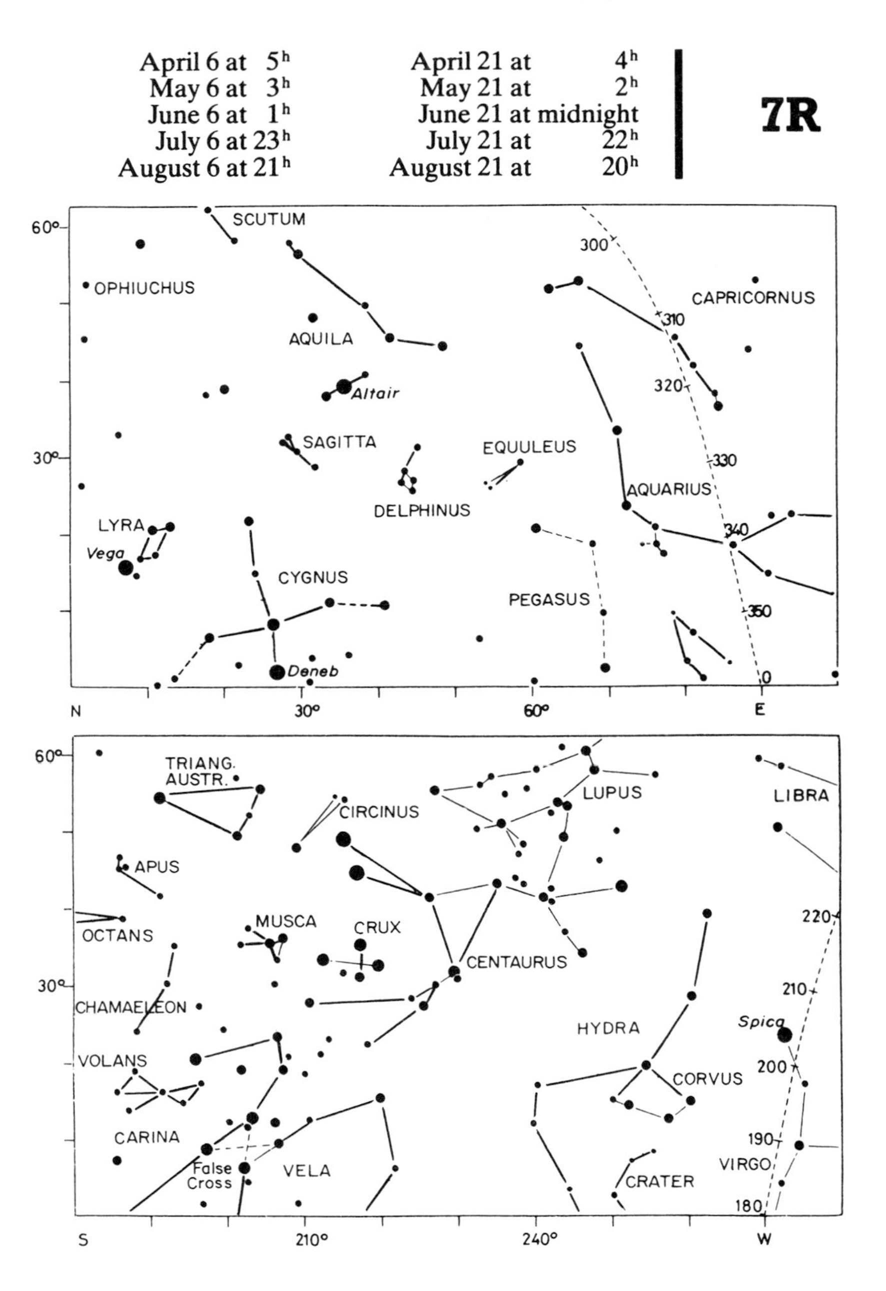
April 6 at 5h
May 6 at 3h
June 6 at 1h
July 6 at 23h
August 6 at 21h
April 21 at 4h
May 21 at 2h
June 21 at midnight
July 21 at 22h
August 21 at 20h
7R
SCUTUM
OPHIUCHUS
AQUILA
Altair
SAGITTA
DELPHINUS
EQUULEUS
CAPRICORNUS
AQUARIUS
LYRA
Vega
CYGNUS
Deneb
PEGASUS
300
310
320
330
340
350
0
60°
30°
N
30°
60°
E
TRIANG. AUSTR.
CIRCINUS
LUPUS
LIBRA
APUS
OCTANS
MUSCA
CRUX
CENTAURUS
CHAMAELEON
HYDRA
Spica
VOLANS
CORVUS
CARINA
False Cross
VELA
CRATER
VIRGO
220
210
200
190
180
60°
30°
S
210°
240°
W

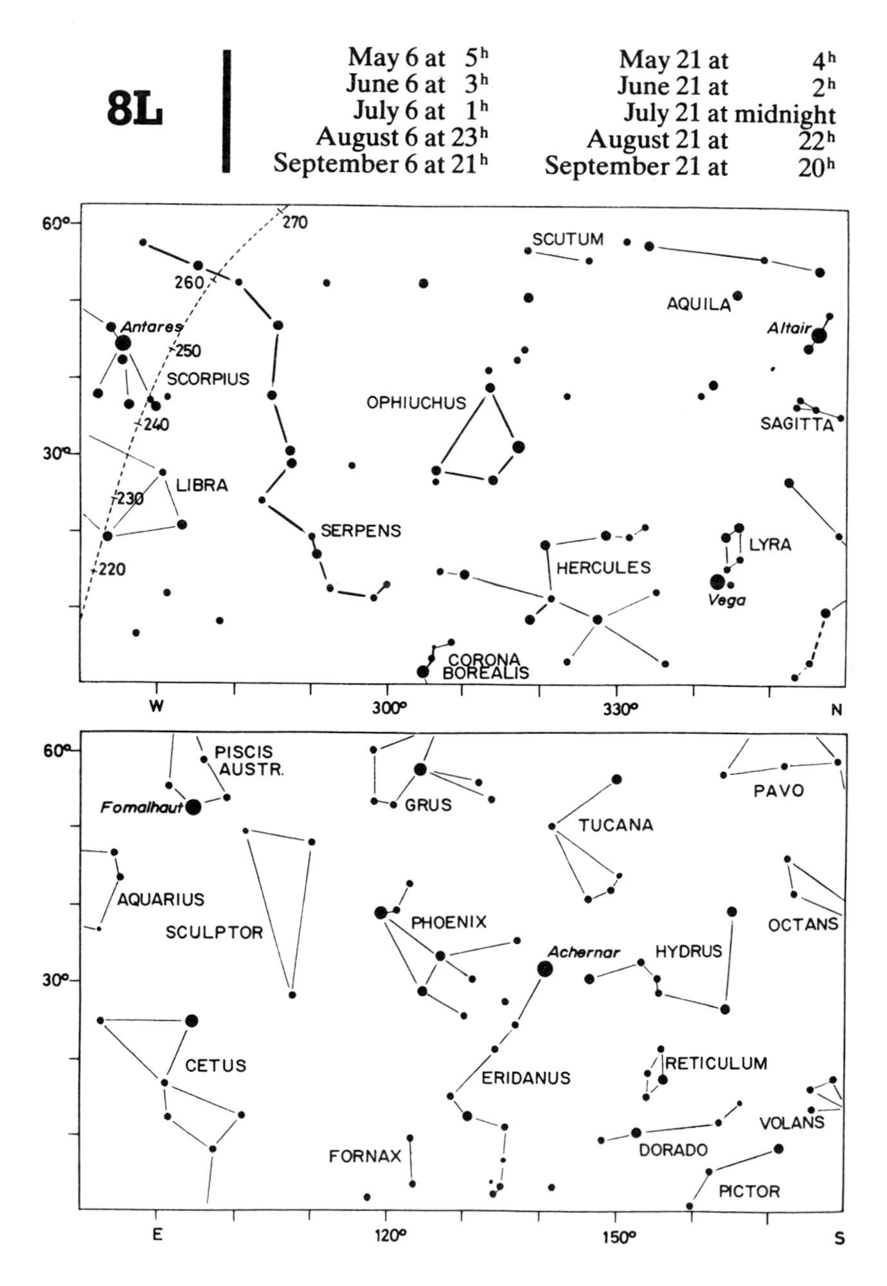
8L
May 6 at 5h
June 6 at 3h
July 6 at 1h
August 6 at 23h
September 6 at 21h
May 21 at 4h
June 21 at 2h
July 21 at midnight
August 21 at 22h
September 21 at 20h
60°
30°
270
260
250
240
230
220
SCUTUM
AQUILA
Altair
Antares
SCORPIUS
OPHIUCHUS
SAGITTA
LIBRA
SERPENS
LYRA
HERCULES
Vega
CORONA
BOREALIS
W
300°
330°
N
PISCIS
AUSTR.
Fomalhaut
GRUS
TUCANA
PAVO
AQUARIUS
SCULPTOR
PHOENIX
OCTANS
Achernar
HYDRUS
CETUS
RETICULUM
ERIDANUS
VOLANS
FORNAX
DORADO
PICTOR
E
120°
150°
S

May 6 at 5^h	May 21 at 4^h	**8R**
June 6 at 3^h	June 21 at 2^h	
July 6 at 1^h	July 21 at midnight	
August 6 at 23^h	August 21 at 22^h	
September 6 at 21^h	September 21 at 20^h	

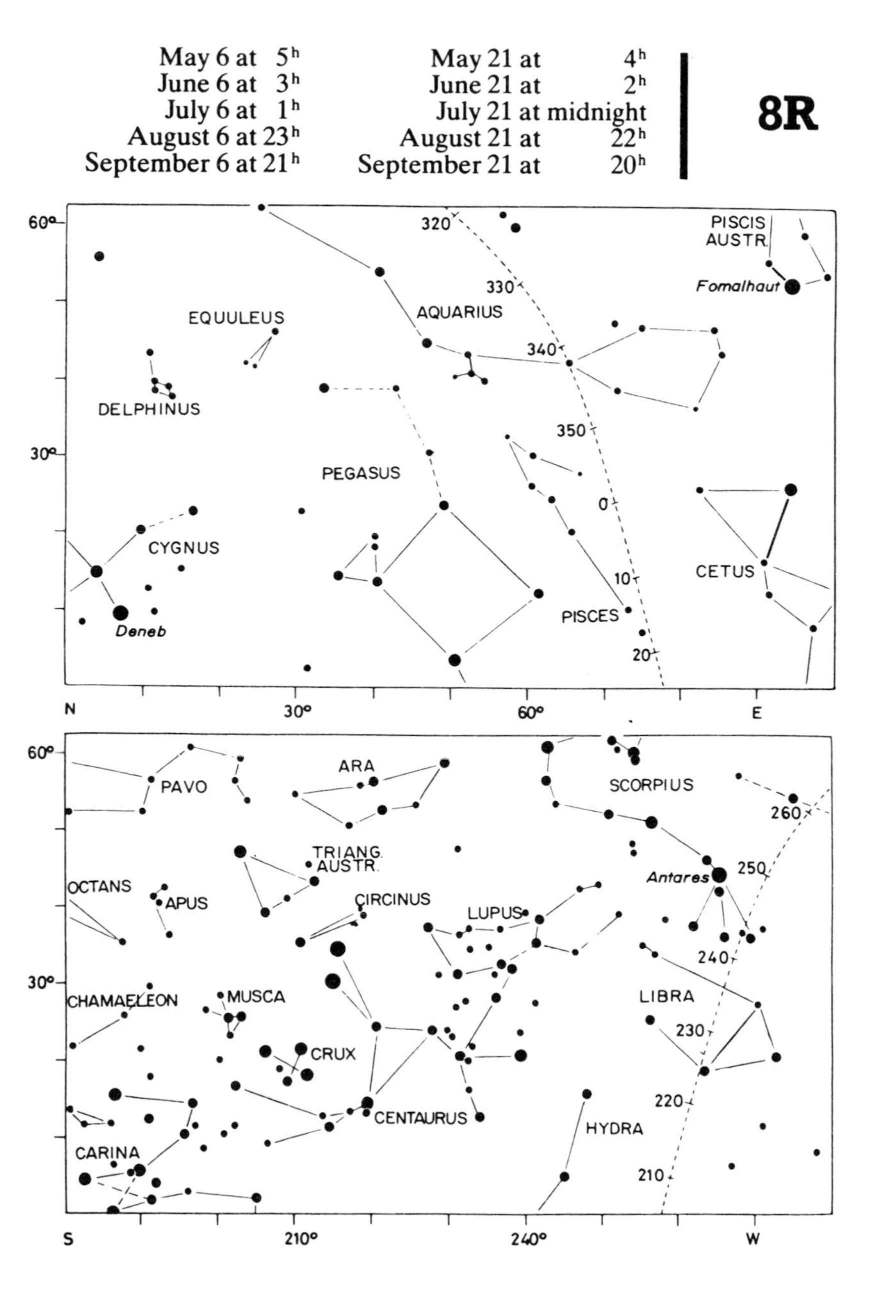

9L

June 6 at 5^h	June 21 at 4^h
July 6 at 3^h	July 21 at 2^h
August 6 at 1^h	August 21 at midnight
September 6 at 23^h	September 21 at 22^h
October 6 at 21^h	October 21 at 20^h

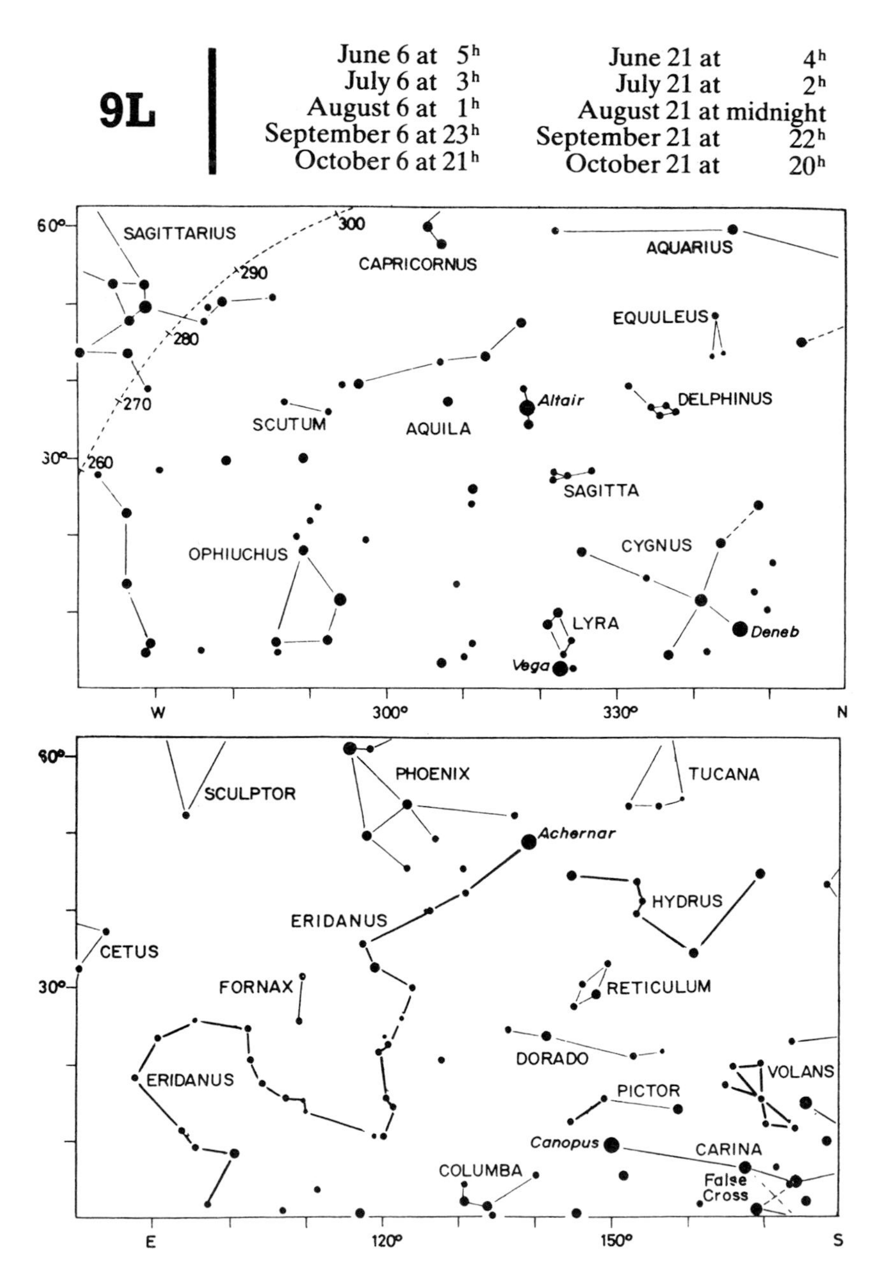

June 6 at 5h	June 21 at 4h	
July 6 at 3h	July 21 at 2h	
August 6 at 1h	August 21 at midnight	**9R**
September 6 at 23h	September 21 at 22h	
October 6 at 21h	October 21 at 20h	

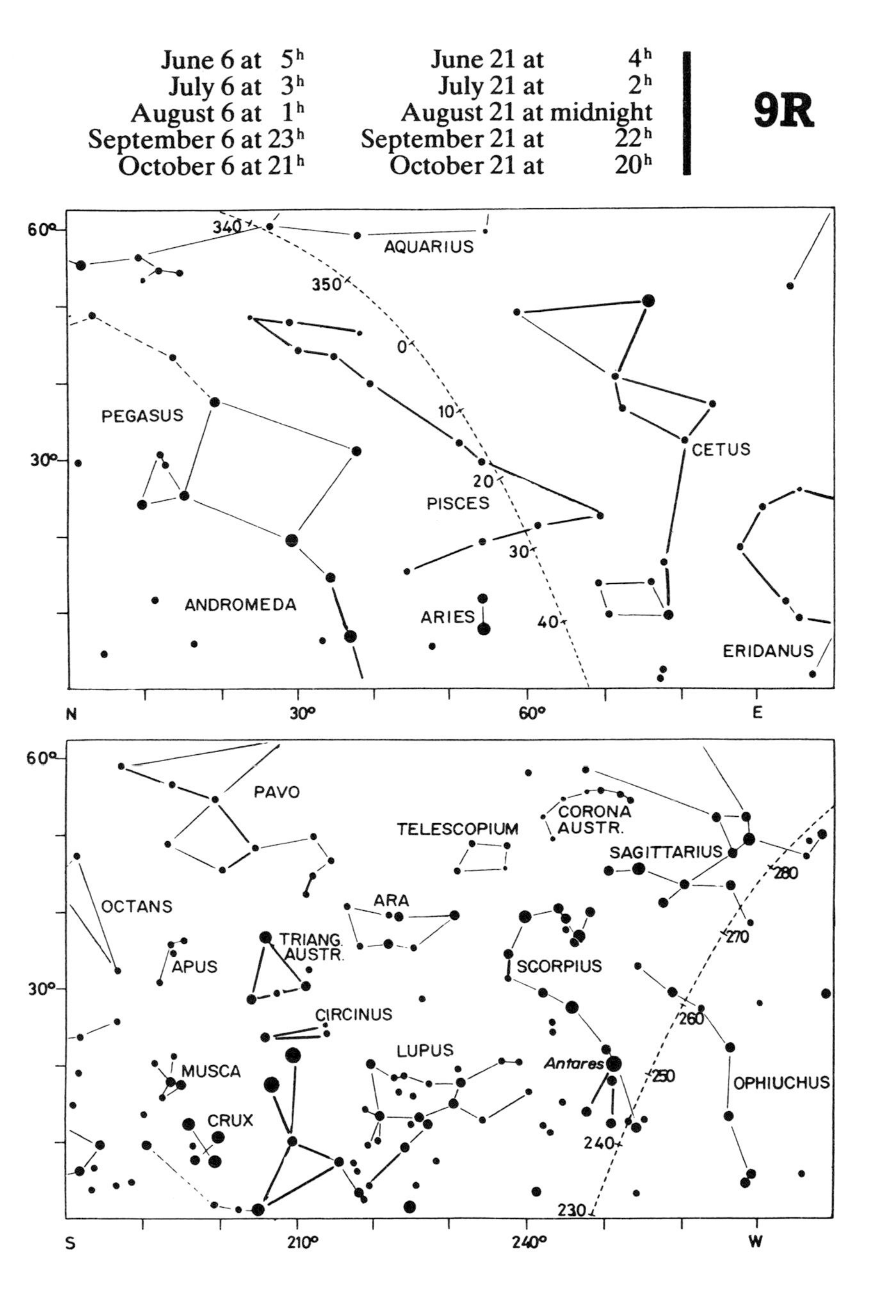

10L

July 6 at 5^h	July 21 at 4^h
August 6 at 3^h	August 21 at 2^h
September 6 at 1^h	September 21 at midnight
October 6 at 23^h	October 21 at 22^h
November 6 at 21^h	November 21 at 20^h

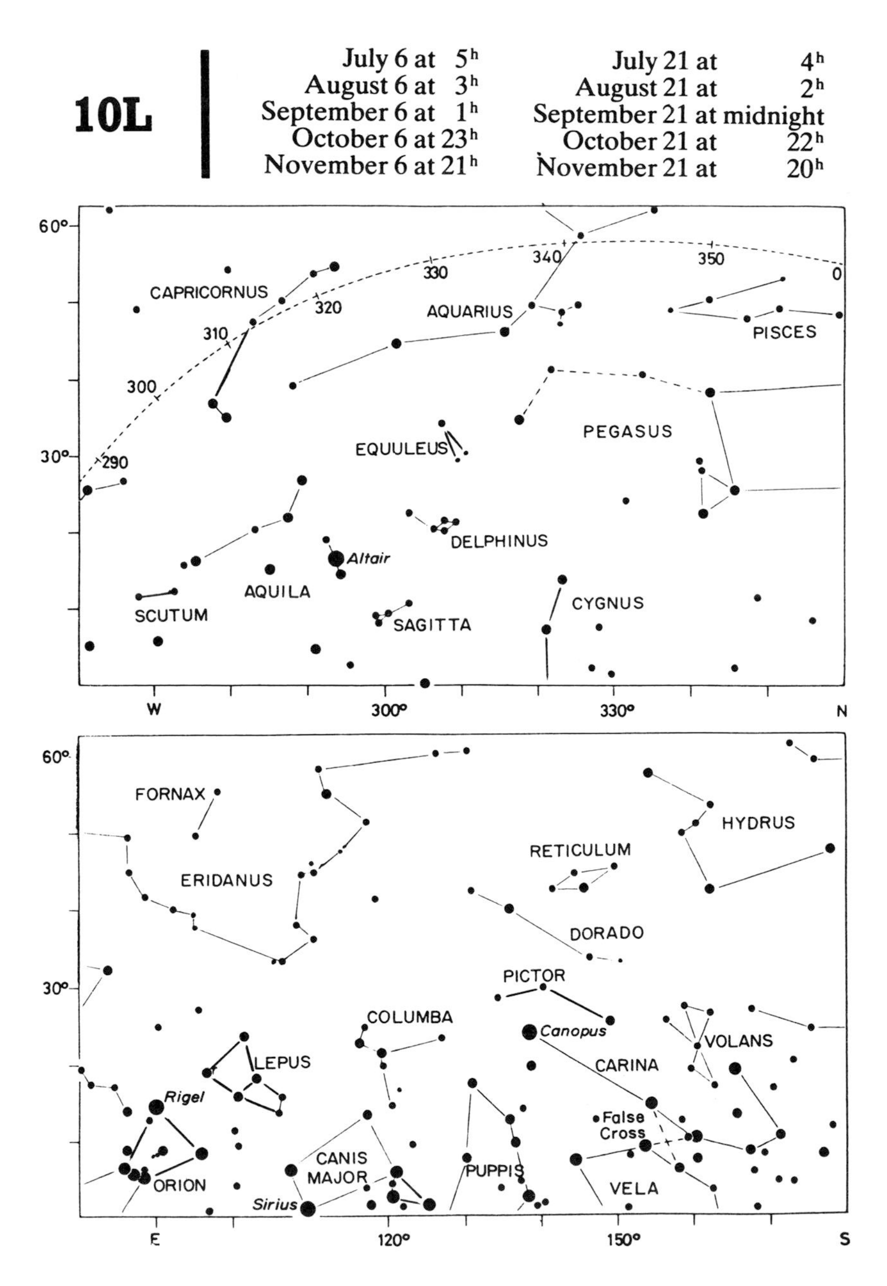

July 6 at 5h	July 21 at 4h	**10R**
August 6 at 3h	August 21 at 2h	
September 6 at 1h	September 21 at midnight	
October 6 at 23h	October 21 at 22h	
November 6 at 21h	November 21 at 20h	

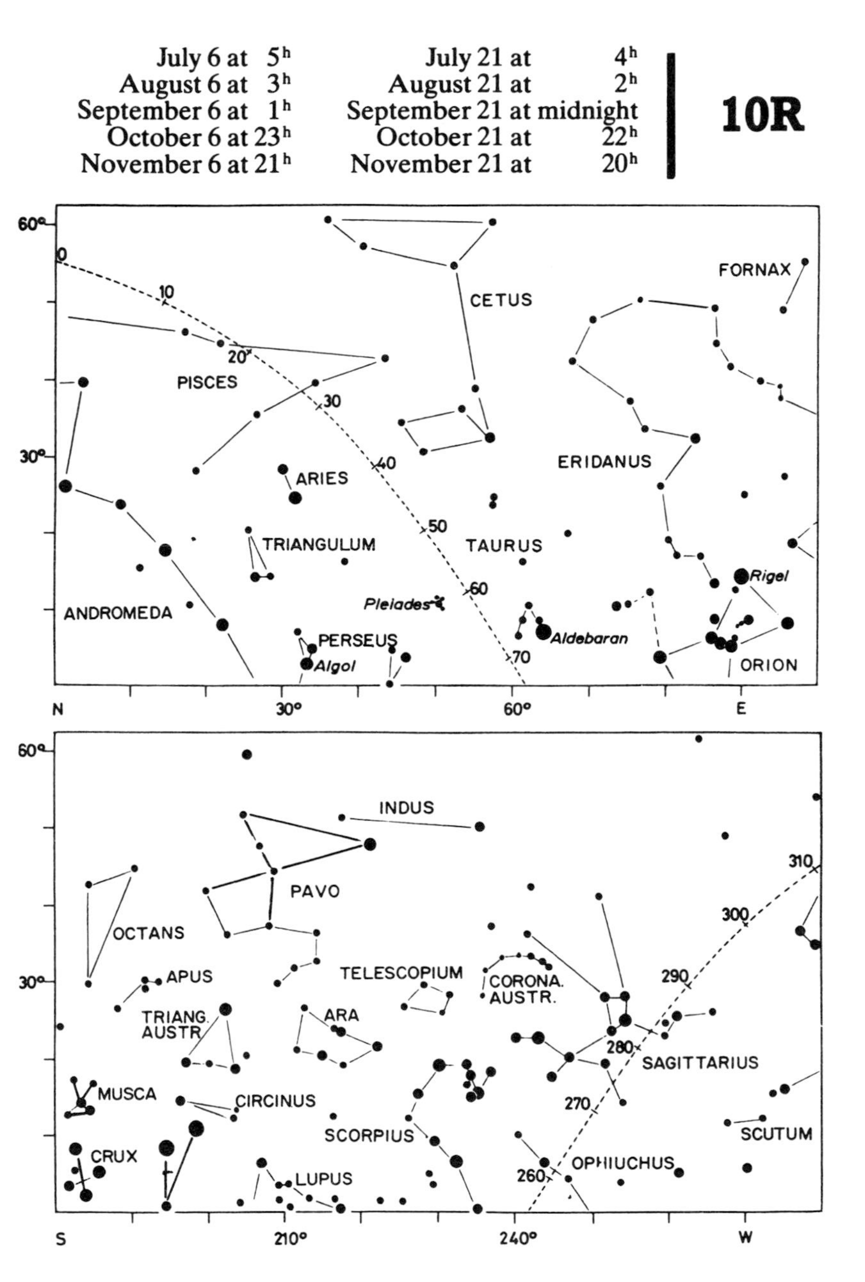

11L

August 6 at 5^h	August 21 at 4^h
September 6 at 3^h	September 21 at 2^h
October 6 at 1^h	October 21 at midnight
November 6 at 23^h	November 21 at 22^h
December 6 at 21^h	December 21 at 20^h

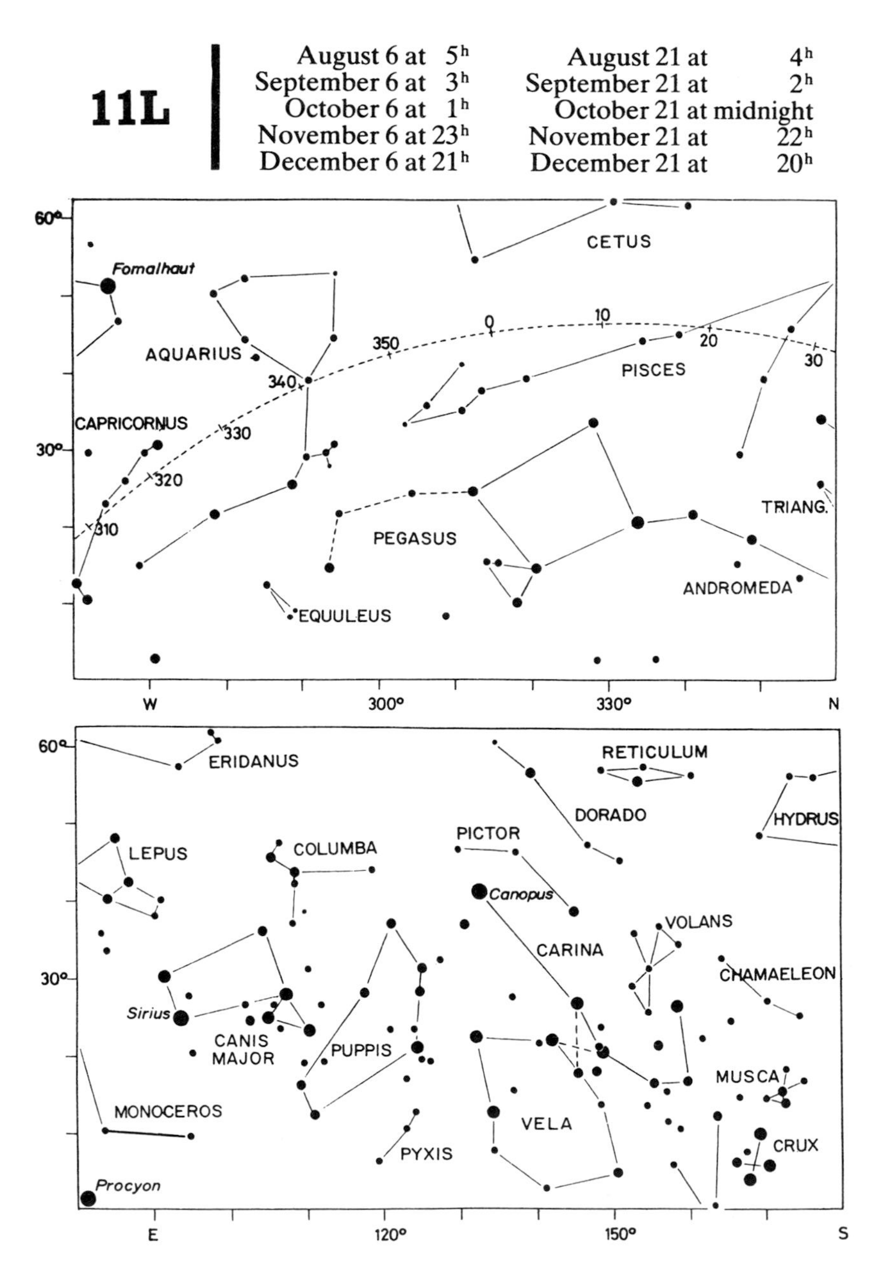

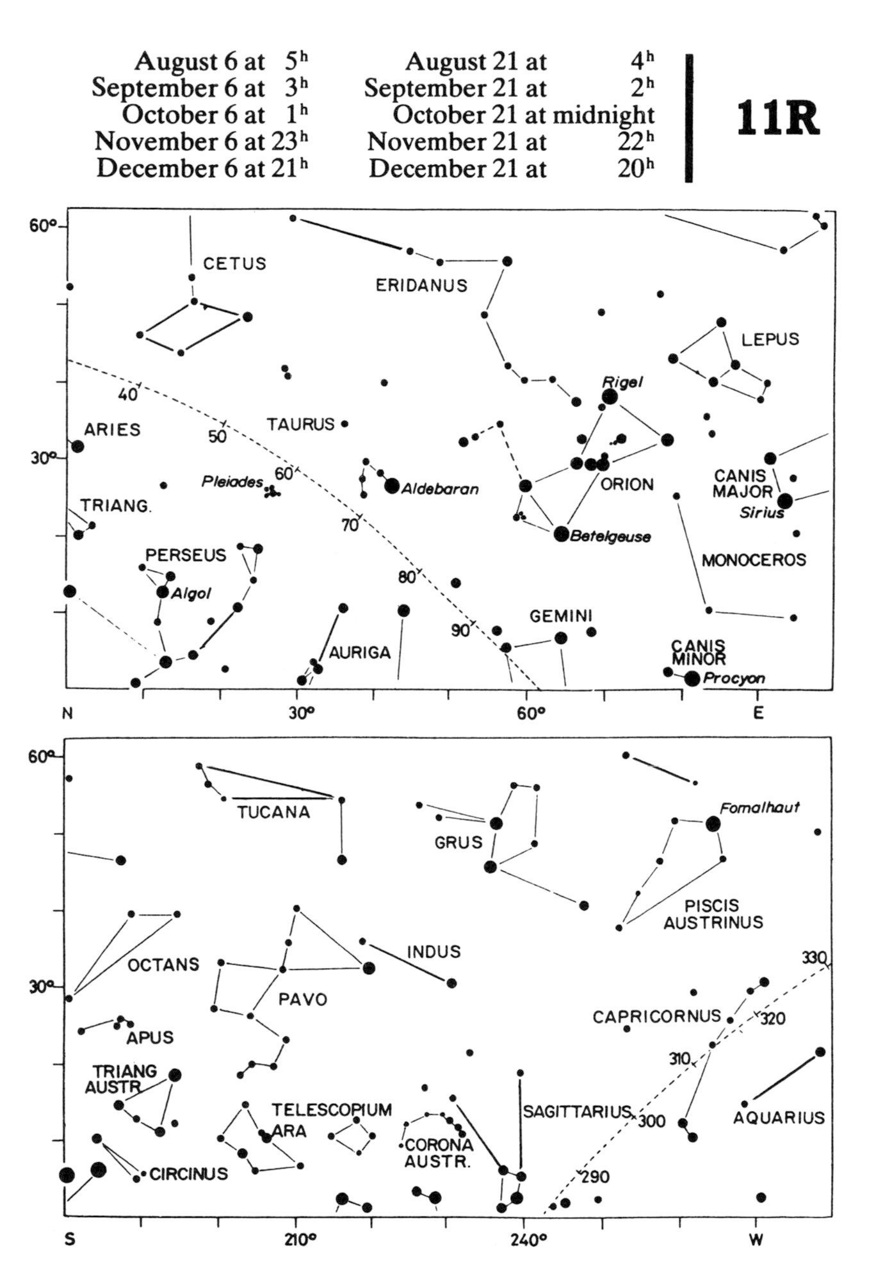
August 6 at 5h
September 6 at 3h
October 6 at 1h
November 6 at 23h
December 6 at 21h
August 21 at 4h
September 21 at 2h
October 21 at midnight
November 21 at 22h
December 21 at 20h
11R
60°
30°
CETUS
ERIDANUS
LEPUS
Rigel
ARIES
TAURUS
ORION
CANIS
MAJOR
Sirius
Pleiades
Aldebaran
TRIANG.
Betelgeuse
PERSEUS
MONOCEROS
Algol
GEMINI
AURIGA
CANIS
MINOR
Procyon
40
50
60
70
80
90
N
30°
60°
E
60°
30°
TUCANA
Fomalhaut
GRUS
PISCIS
AUSTRINUS
INDUS
OCTANS
330
PAVO
CAPRICORNUS
320
APUS
310
TRIANG
AUSTR
TELESCOPIUM
SAGITTARIUS
300
AQUARIUS
ARA
CORONA
AUSTR.
CIRCINUS
290
S
210°
240°
W

12L

September 6 at 5ʰ	September 21 at 4ʰ
October 6 at 3ʰ	October 21 at 2ʰ
November 6 at 1ʰ	November 21 at midnight
December 6 at 23ʰ	December 21 at 22ʰ
January 6 at 21ʰ	January 21 at 20ʰ

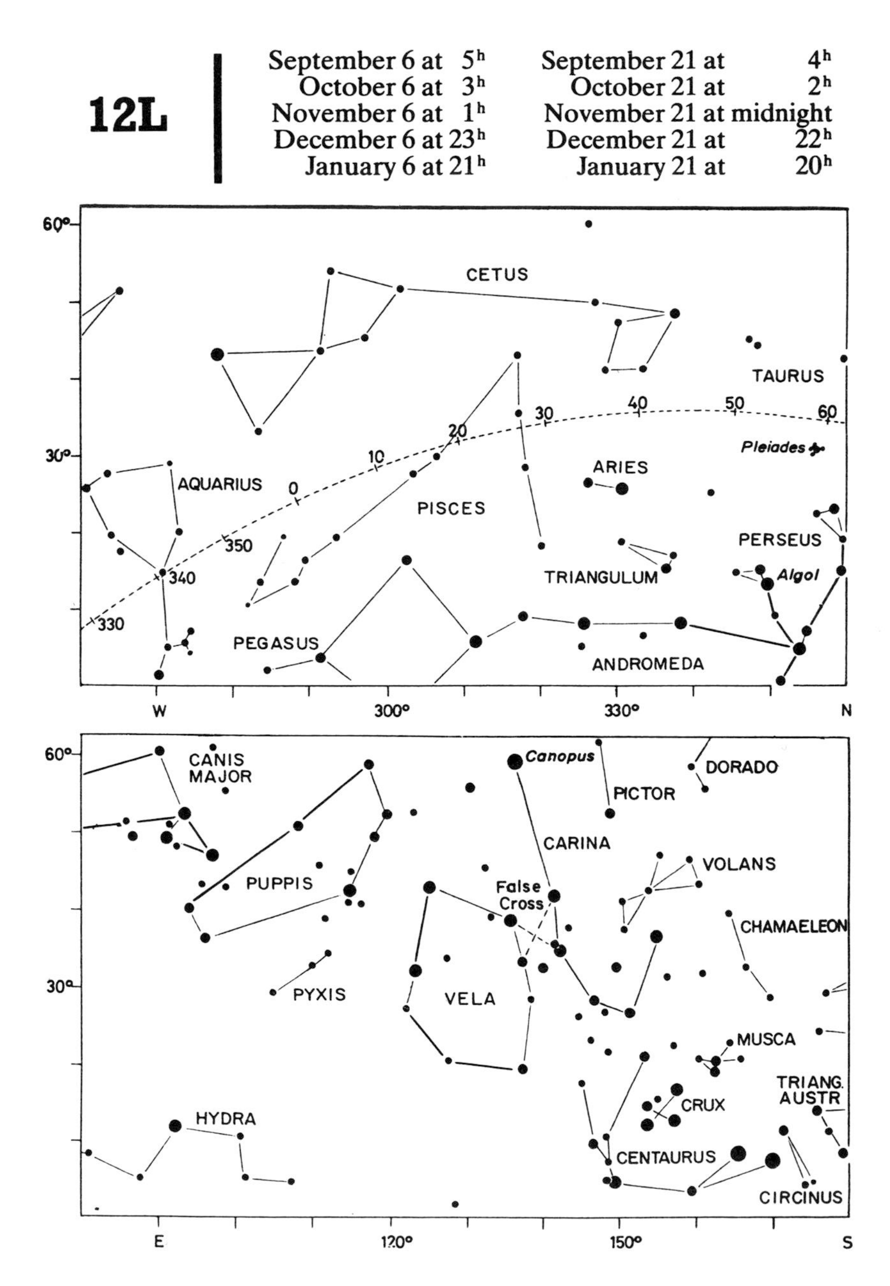

September 6 at 5h	September 21 at 4h	**12R**
October 6 at 3h	October 21 at 2h	
November 6 at 1h	November 21 at midnight	
December 6 at 23h	December 21 at 22h	
January 6 at 21h	January 21 at 20h	

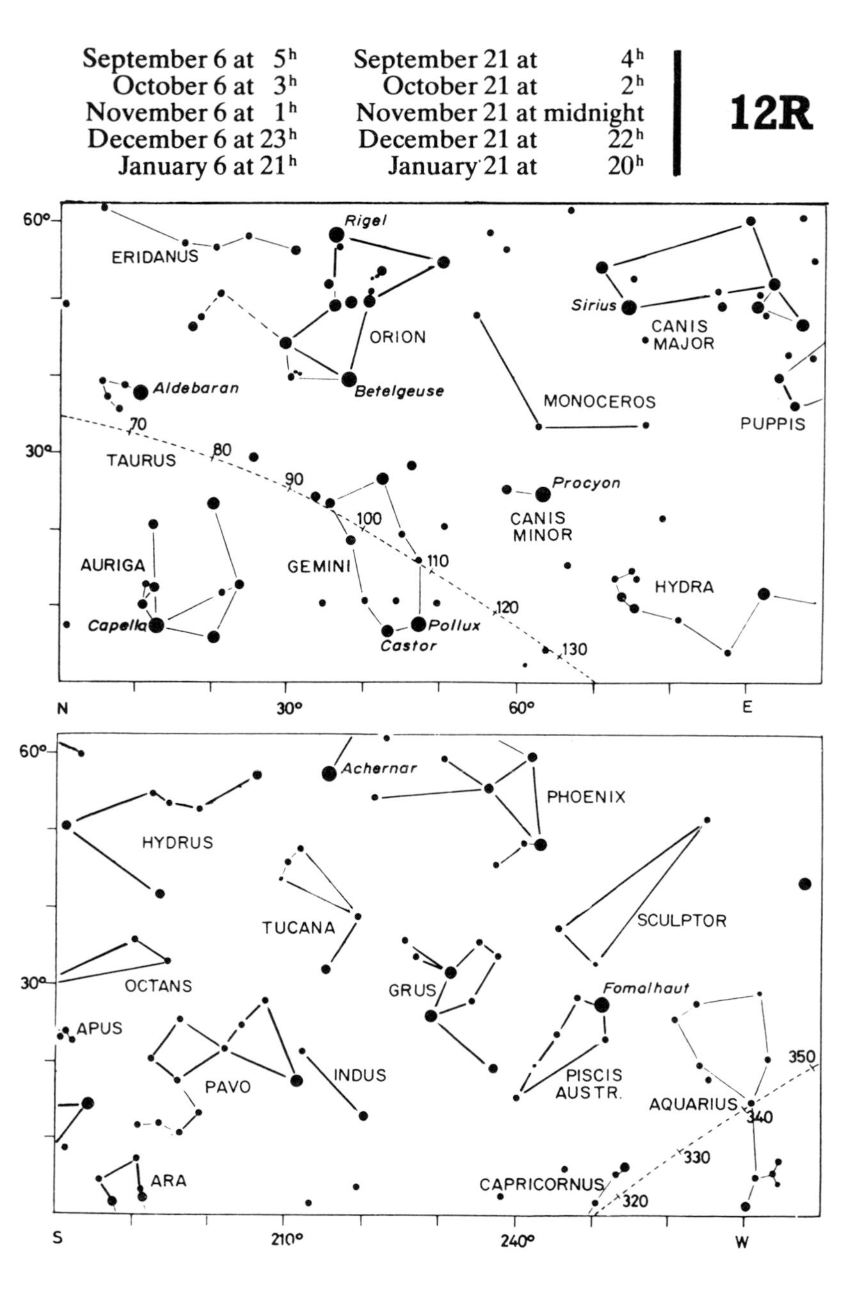

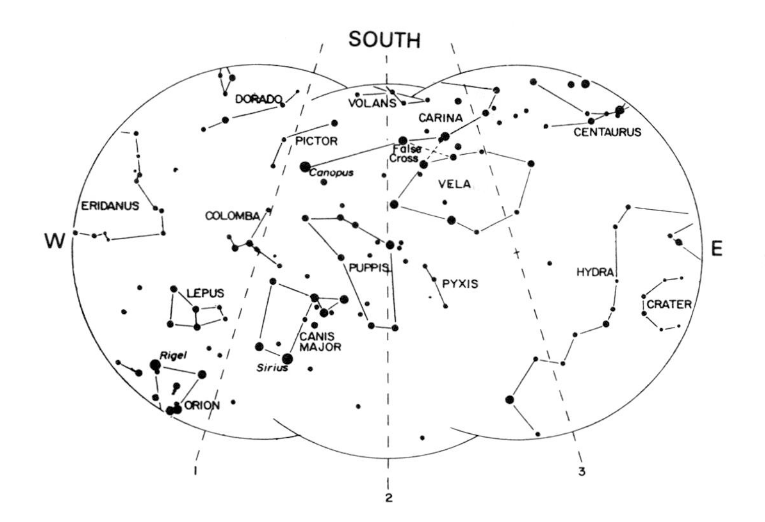

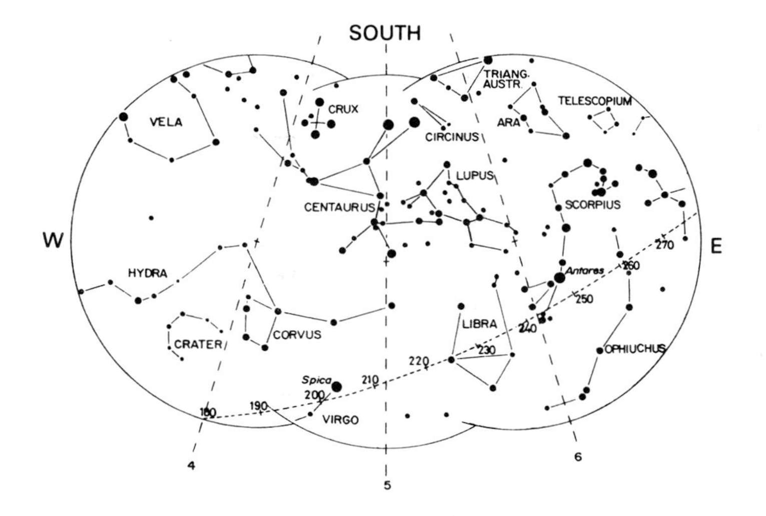

Southern Hemisphere Overhead Stars

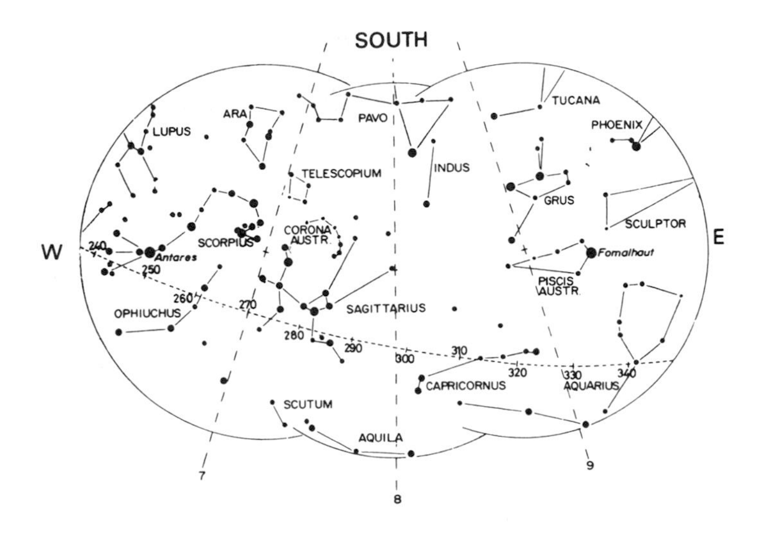

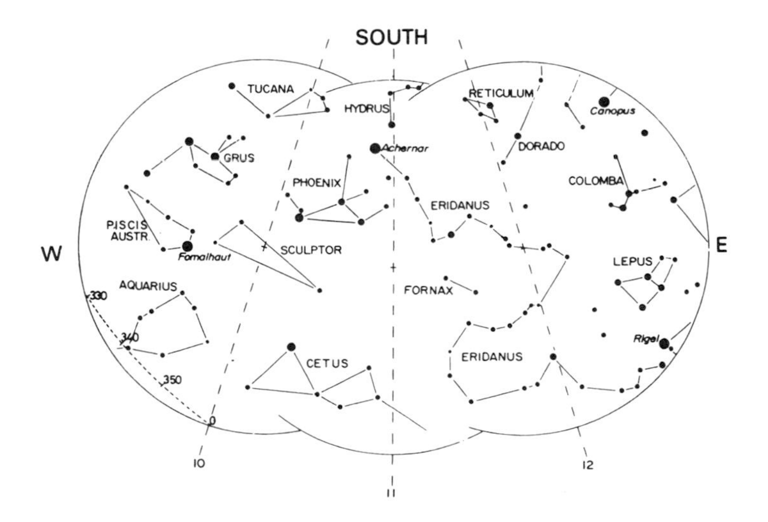

Southern Hemisphere Overhead Stars

The Planets and the Ecliptic

The paths of the planets about the Sun all lie close to the plane of the ecliptic, which is marked for us in the sky by the apparent path of the Sun among the stars, and is shown on the star charts by a broken line. The Moon and planets will always be found close to this line, never departing from it by more than about 7 degrees. Thus the planets are most favourably placed for observation when the ecliptic is well displayed, and this means that it should be as high in the sky as possible. This avoids the difficulty of finding a clear horizon, and also overcomes the problem of atmospheric absorption, which greatly reduces the light of the stars. Thus a star at an altitude of 10 degrees suffers a loss of 60 per cent of its light, which corresponds to a whole magnitude; at an altitude of only 4 degrees, the loss may amount to two magnitudes.

The position of the ecliptic in the sky is therefore of great importance, and since it is tilted at about 23½ degrees to the Equator, it is only at certain times of the day or year that it is displayed to the best advantage. It will be realized that the Sun (and therefore the ecliptic) is at its highest in the sky at noon in midsummer, and at its lowest at noon in midwinter. Allowing for the daily motion of the sky, these times lead to the fact that the ecliptic is highest at midnight in winter, at sunset in the spring, at noon in summer and at sunrise in the autumn. Hence these are the best times to see the planets. Thus, if Venus is an evening object in the western sky after sunset, it will be seen to best advantage if this occurs in the spring, when the ecliptic is high in the sky and slopes down steeply to the horizon. This means that the planet is not only higher in the sky, but will remain for a much longer period above the horizon. For similar reasons, a morning object will be seen at its best on autumn mornings before sunrise, when the ecliptic is high in the east. The outer planets, which can come to opposition (i.e. opposite the Sun), are best seen when opposition occurs in the winter months, when the ecliptic is high in the sky at midnight.

The seasons are reversed in the Southern Hemisphere, spring beginning at the September Equinox, when the Sun crosses the Equator on its way south, summer beginning at the December

Solstice, when the Sun is highest in the southern sky, and so on. Thus, the times when the ecliptic is highest in the sky, and therefore best placed for observing the planets, may be summarized as follows:

	Midnight	*Sunrise*	*Noon*	*Sunset*
Northern lats.	December	September	June	March
Southern lats.	June	March	December	September

In addition to the daily rotation of the celestial sphere from east to west, the planets have a motion of their own among the stars. The apparent movement is generally *direct*, i.e. to the east, in the direction of increasing longitude, but for a certain period (which depends on the distance of the planet) this apparent motion is reversed. With the outer planets this *retrograde* motion occurs about the time of opposition. Owing to the different inclination of the orbits of these planets, the actual effect is to cause the apparent path to form a loop, or sometimes an S-shaped curve. The same effect is present in the motion of the inferior planets, Mercury and Venus, but it is not so obvious, since it always occurs at the time of inferior conjunction.

The inferior planets, Mercury and Venus, move in smaller orbits than that of the Earth, and so are always seen near the Sun. They are most obvious at the times of greatest angular distance from the Sun (greatest elongation), which may reach 28 degrees for Mercury, or 47 degrees for Venus. They are seen as evening objects in the western sky after sunset (at eastern elongations) or as morning objects in the eastern sky before sunrise (at western elongations). The succession of phenomena, conjunctions and elongations always follows the same order, but the intervals between them are not equal. Thus, if either planet is moving round the far side of its orbit its motion will be to the east, in the same direction in which the Sun appears to be moving. It therefore takes much longer for the planet to overtake the Sun – that is, to come to superior conjunction – than it does when moving round to inferior conjunction, between Sun and Earth. The intervals given in the following table are average values; they remain fairly constant in the case of Venus, which travels in an almost circular orbit. In the case of Mercury, however, conditions vary widely because of the great eccentricity and inclination of the planet's orbit.

		Mercury	*Venus*
Inferior conj.	to Elongation West	22 days	72 days
Elongation West	to Superior conj.	36 days	220 days
Superior conj.	to Elongation East	36 days	220 days
Elongation East	to Inferior conj.	22 days	72 days

The greatest brilliancy of Venus always occurs about 36 days before or after inferior conjunction. This will be about a month *after* greatest eastern elongation (as an evening object), or a month *before* greatest western elongation (as a morning object). No such rule can be given for Mercury, because its distance from the Earth and the Sun can vary over a wide range.

Mercury is not likely to be seen unless a clear horizon is available. It is seldom seen as much as 10 degrees above the horizon in the twilight sky in northern latitudes, but this figure is often exceeded in the Southern Hemisphere. This favourable condition arises because the maximum elongation of 28 degrees can occur only when the planet is at aphelion (farthest from the Sun), and this point lies well south of the Equator. Northern observers must be content with smaller elongations, which may be as little as 18 degrees at perihelion. In general, it may be said that the most favourable times for seeing Mercury as an evening object will be in spring, some days before greatest eastern elongation; in autumn, it may be seen as a morning object some days after greatest western elongation.

Venus is the brightest of the planets and may be seen on occasions in broad daylight. Like Mercury, it is alternately a morning and an evening object, and it will be highest in the sky when it is a morning object in autumn, or an evening object in spring. The phenomena of Venus given in the table above can occur only in the months of January, April, June, August and November, and it will be realized that they do not all lead to favourable apparitions of the planet. In fact, Venus is to be seen at its best as an evening object in northern latitudes when eastern elongation occurs in June. The planet is then well north of the Sun in the preceding spring months, and is a brilliant object in the evening sky over a long period. In the Southern Hemisphere a November elongation is best. For similar reasons, Venus gives a prolonged display as a morning object in the months following western elongation in November (in northern latitudes) or in June (in the Southern Hemisphere).

The superior planets, which travel in orbits larger than that of the Earth, differ from Mercury and Venus in that they can be seen opposite the Sun in the sky. The superior planets are morning objects after conjunction with the Sun, rising earlier each day until they come to opposition. They will then be nearest to the Earth (and therefore at their brightest), and will then be on the meridian at midnight, due south in northern latitudes, but due north in the Southern Hemisphere. After opposition they are evening objects, setting earlier each evening until they set in the west with the Sun at the next conjunction. The change in brightness about the time of opposition is most noticeable in the case of Mars, whose distance from Earth can vary considerably and rapidly. The other superior planets are at such great distances that there is very little change in brightness from one opposition to another. The effect of altitude is, however, of some importance, for at a December opposition in northern latitudes the planets will be among the stars of Taurus or Gemini, and can then be at an altitude of more than 60 degrees in southern England. At a summer opposition, when the planet is in Sagittarius, it may only rise to about 15 degrees above the southern horizon, and so makes a less impressive appearance. In the Southern Hemisphere, the reverse conditions apply, a June opposition being the best, with the planet in Sagittarius at an altitude which can reach 80 degrees above the northern horizon for observers in South Africa.

Mars, whose orbit is appreciably eccentric, comes nearest to the Earth at an opposition at the end of August. It may then be brighter even than Jupiter, but rather low in the sky in Aquarius for northern observers, though very well placed for those in southern latitudes. These favourable oppositions occur every fifteen or seventeen years (1971, 1988, 2003, 2018) but in the Northern Hemisphere the planet is probably better seen at an opposition in the autumn or winter months, when it is higher in the sky. Oppositions of Mars occur at an average interval of 780 days, and during this time the planet makes a complete circuit of the sky.

Jupiter is always a bright planet, and comes to opposition a month later each year, having moved, roughly speaking, from one Zodiacal constellation to the next.

Saturn moves much more slowly than Jupiter, and may remain in the same constellation for several years. The brightness of Saturn depends on the aspects of its rings, as well as on the distance from Earth and Sun. The rings were inclined towards the Earth and Sun in 1995 and are now past their maximum opening. The next passage of both Earth and Sun through the ring-plane will not occur until 2009.

Uranus, *Neptune*, and *Pluto* are hardly likely to attract the attention of observers without adequate instruments.

Phases of the Moon, 1996

New Moon				*First Quarter*				*Full Moon*				*Last Quarter*			
	d	h	m		d	h	m		d	h	m		d	h	m
Jan.	20	12	50	Jan.	27	11	14	Jan.	5	20	51	Jan.	13	20	45
Feb.	18	23	30	Feb.	26	05	52	Feb.	4	15	58	Feb.	12	08	37
Mar.	19	10	45	Mar.	27	01	31	Mar.	5	09	23	Mar.	12	17	15
Apr.	17	22	49	Apr.	25	20	40	Apr.	4	00	07	Apr.	10	23	36
May	17	11	46	May	25	14	13	May	3	11	48	May	10	05	04
June	16	01	36	June	24	05	23	June	1	20	47	June	8	11	06
July	15	16	15	July	23	17	49	July	1	03	58	July	7	18	55
Aug.	14	07	34	Aug.	22	03	36	July	30	10	35	Aug.	6	05	25
Sept.	12	23	07	Sept.	20	11	23	Aug.	28	17	52	Sept.	4	19	06
Oct.	12	14	14	Oct.	19	18	09	Sept.	27	02	51	Oct.	4	12	04
Nov.	11	04	16	Nov.	18	01	09	Oct.	26	14	11	Nov.	3	07	50
Dec.	10	16	56	Dec.	17	09	31	Nov.	25	04	10	Dec.	3	05	06
								Dec.	24	20	41				

All times are GMT.

Longitudes of the Sun, Moon and Planets in 1996

DATE		*Sun*	*Moon*	*Venus*	*Mars*	*Jupiter*	*Saturn*
		°	°	°	°	°	°
January	6	285	106	319	298	271	350
	21	300	307	336	309	274	351
February	6	316	151	356	322	277	353
	21	332	359	13	334	280	354
March	6	346	173	30	345	283	356
	21	1	21	46	357	285	358
April	6	16	221	62	10	286	0
	21	31	67	75	21	287	2
May	6	46	259	85	32	288	3
	21	60	100	88	43	287	5
June	6	76	313	83	55	286	6
	21	90	144	74	66	285	7
July	6	104	351	72	76	283	7
	21	119	177	77	86	281	7
August	6	134	41	89	98	279	7
	21	148	224	102	107	278	7
September	6	164	87	119	118	278	6
	21	178	275	134	127	278	4
October	6	193	119	152	136	279	3
	21	208	314	169	144	281	2
November	6	224	163	189	154	284	1
	21	239	7	207	161	286	1
December	6	254	195	226	169	289	1
	21	269	44	244	175	293	1

Longitude of *Uranus* 302°
Neptune 296°

Moon: Longitude of ascending node
Jan. 1: 202° Dec. 31: 183°

Mercury moves so quickly among the stars that it is not possible to indicate its position on the star charts at a convenient interval. The monthly notes must be consulted for the best times at which the planet may be seen.

The positions of the other planets are given in the table on the previous page. This gives the apparent longitudes on dates which correspond to those of the star charts, and the position of the planet may at once be found near the ecliptic at the given longitude.

Examples

In the Northern Hemisphere two planets are seen in the western evening sky in early February. Identify them.

> The Northern Star Chart 12R shows the western sky at February 6^{d} 19^{h} and shows longitudes 340°–60°. Reference to the table on page 71 gives the longitude of Venus as 356° and that of Saturn as 353°. Thus these planets are to be found in the eastern sky and the one lower down is Saturn.

The positions of the Sun and Moon can be plotted on the star maps in the same manner as for the planets. The average daily motion of the Sun is 1°, and of the Moon 13°. For the Moon an indication of its position relative to the ecliptic may be obtained from a consideration of its longitude relative to that of the ascending node. The latter changes only slowly during the year as will be seen from the values given on the previous page. Let us call the difference in longitude of Moon-node, d. Then if d = 0°, 180° or 360° the Moon is on the ecliptic. If d = 90° the Moon is 5° north of the ecliptic and if d = 270° the Moon is 5° south of the ecliptic.

On May 21 the Moon's longitude is given as 100° and the longitude of the node is found by interpolation to be about 194°. Thus d = 266° and the Moon is about 5° south of the ecliptic. Its position may be plotted on Northern Star Charts 1L, 2R, 3R, 4L, 4R, 5L, 10R, 11L, 11R, and 12L, and Southern Star Charts 1R, 2L, 3L, and 12R.

Events in 1996

ECLIPSES

There will be four eclipses, two of the Sun and two of the Moon.

April 3–4: total eclipse of the Moon – western Asia, Europe, Americas.
April 17: partial eclipse of the Sun – New Zealand.
September 27: total eclipse of the Moon – western Asia, Europe, Americas.
October 12: partial eclipse of the Sun – Canada, Europe, N. Africa.

THE PLANETS

Mercury may be seen more easily from northern latitudes in the evenings about the time of greatest eastern elongation (April 23) and in the mornings around greatest western elongation (October 3). In the Northern Hemisphere the corresponding dates are February 11 (morning) and August 21 (evening).

Venus is visible in the evenings until May and in the mornings from mid-June to December.

Mars is visible in the mornings from June to December.

Jupiter is at opposition on July 4.

Saturn is at opposition on September 26.

Uranus is at opposition on July 25.

Neptune is at opposition on July 18.

Pluto is at opposition on May 22.

JANUARY

New Moon: January 20 *Full Moon:* January 5

EARTH is at perihelion (nearest to the Sun) on January 4 at a distance of 147 million kilometres.

MERCURY is at greatest eastern elongation (19°) on January 2. For the first ten days of the month it may be possible for keen sighted observers to detect it low above the south-western horizon, about half-an-hour after sunset. During this period its magnitude fades from −0.6 to +0.5. It passes through inferior conjunction on January 18. For the last week of the month observers (though only those in southern latitudes) will be able to see Mercury in the mornings about half-an-hour before dawn, low above the east-south-east horizon. They should refer to Figure 2 in the notes for February.

VENUS is a brilliant evening object in the south-western sky, magnitude −4.0. Observers in the British Isles will note that the period available for observation increases from two to three hours during the month as the planet moves northwards in declination and increases its angular distance from the Sun.

MARS is too close to the Sun for observation.

JUPITER, magnitude −1.8, is emerging from the morning twilight, becoming visible about the south-eastern horizon before dawn. Observers in equatorial and southern latitudes will be able to see the planet quite early in the month but because of its southern declination those in northern temperate latitudes will not be able to see it before the middle of the month. For these observers it will never be very high above the horizon, even when on the meridian, since Jupiter remains in Sagittarius throughout 1996.

SATURN, magnitude +1.2, is an evening object in the south-western sky. Saturn is in Aquarius and its path amongst the stars is shown in Figure 8, given with the notes for September.

FINDING THE SOUTH POLE STAR. Northern navigators have always been glad to have a bright pole star – Polaris in Ursa Minor (the Little Bear), which is of the second magnitude and is very easy to find. Things are less easy in the south, where Polaris remains below the horizon; there is no bright south pole star, and the nearest which is reasonably prominent (Beta Hydri, magnitude 2.8) is well over ten degrees from the polar point. So we have to make do with the obscure Sigma Octantis, magnitude 5.5 – which is none too easy to identify even with a clear sky, while the slightest mist or haze will hide it.

Octans, the Octant, is a very barren constellation. The brightest star in it, Nu Octantis, is only of magnitude 3.8. Neither is there any definite pattern. The pole lies about midway between the Southern Cross and the brilliant Achernar in Eridanus, but to identify Sigma Octantis something more detailed is required.

One method is to start with Alpha Centauri, the brighter of the two Pointers to the Southern Cross. Close beside it is Alpha Circini, magnitude 3.2. These show the way to the little constellation of Apus; Alpha Apodis is of magnitude 3.8. In the same binocular field as Alpha Apodis are two dim stars, Epsilon Apodis (5.2) and Eta Apodis (5.0); these point straight to the orange Delta Octantis (4.3).

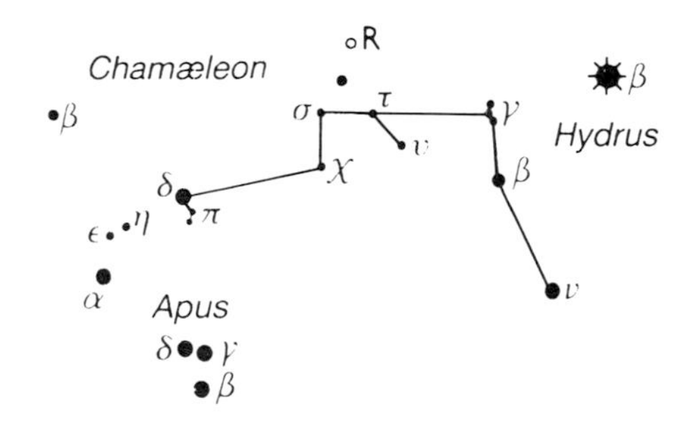

Figure 1. The constellation of Apus.

Delta can be identified because of the two faint stars close beside it. Now put Delta Octantis in the edge of the field, and continue the line from Apus. Chi Octantis (5.2) will be on the far side of the field, centre it, and you will then see two more stars of about the same brightness, Sigma and Tau. These three make up a triangle, in the same field with, say, 7 × 50 binoculars. Sigma is the second star in order from Delta.

This may sound rather complicated, but I have never found an easier method! Moreover, Sigma is almost one degree from the polar point. It is about six times as luminous as the Sun, with a slightly hotter surface; its distance is 120 light-years.

THE ASTRONOMICAL UNIT. The astronomical unit, or Earth–Sun distance, is the mean distance between the two bodies; in round figures, 150,000,000 kilometres or 93,000,000 miles. Early estimates were decidedly wide of the mark. In modern units, the value given by the Greek philosopher Anaxagoras (500–428 BC) was a mere 6500 km. The first reasonably good estimate was due to G. D. Cassini in 1672; using measures of the parallax of Mars, he gave 138,370,000 km. Later estimates have been as follows:

Year	*Authority*	*Method*	*Distance, km*
1770	Euler	Transit of Venus, 1769	151,225,000
1862	Foucault	Velocity of light	147,469,000
1875	Galle	Parallax of the asteroid Flora	148,290,000
1877	Airy	Transit of Venus, 1874	150,152,000
1931	Spencer Jones	Parallax of the asteroid Eros	149,645,000
1992	Various	Radar	149,598,893

The value accepted today is accurate to a tiny fraction of one per cent.

Represent the Earth–Sun distance by one inch, and the nearest star beyond the Sun, Proxima Centauri, will be over 4 miles away (for this comparison Imperial units are more convenient than metric).

FEBRUARY

New Moon: February 18 *Full Moon:* February 4

MERCURY is well south of the Sun and thus poorly placed for observation by those in northern temperate latitudes. For observers in southern latitudes this will be the most favourable morning apparition of the year. Figure 2 shows, for observers in latitude S.35°, the changes in azimuth (true bearing from the north through east, south, and west) and altitude of Mercury on successive evenings when the Sun is 6° below the horizon. This condition is known as the beginning of morning civil twilight and in this latitude and at this time of year occurs about 30 minutes before sunrise. The changes in the brightness of the planet are indicated by the relative sizes of the circles marking Mercury's position at five-day intervals. It will be noticed that Mercury is at its brightest after it reaches greatest western elongation (26°) on February 11.

VENUS continues to be visible as a brilliant evening object, magnitude −4.1. It is visible in the western sky for several hours after sunset. At the beginning of the month Venus and Saturn are near each other and on February 2 Venus passes only 1° south of Saturn.

MARS is in conjunction with the Sun early next month and thus is unsuitably placed for observation.

JUPITER is visible as a morning object in the south-eastern sky before dawn, magnitude −1.9.

SATURN is still an evening object in the south-western sky but coming towards the end of its apparition. Its magnitude is +1.2. Saturn will be a disappointing object for observers since the Earth passes through the ring plane on February 11–12. For observers in the British Isles Saturn will be lost to view in the gathering twilight before the end of the month.

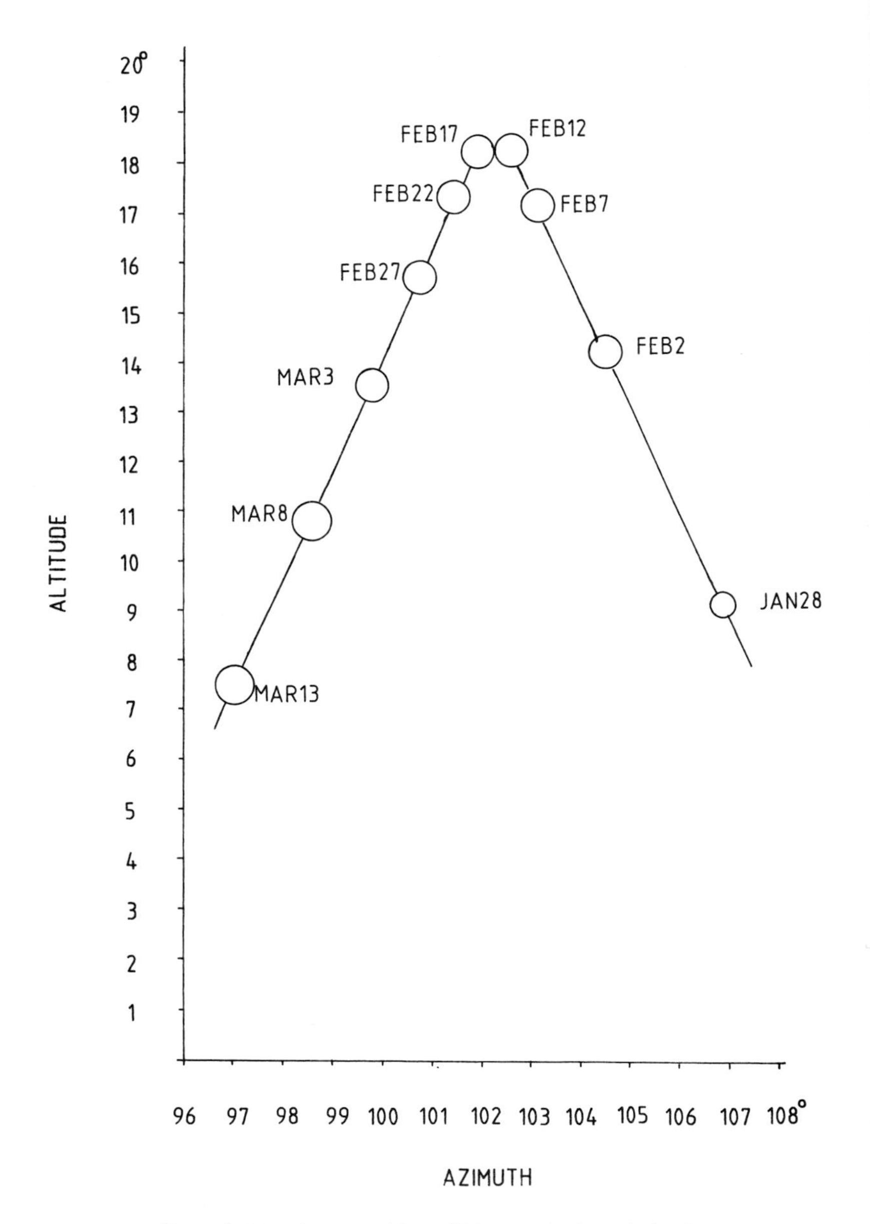

Figure 2. Morning apparition of Mercury for latitude S.35°.

LUDWIG'S STAR. One of the most famous stars in the sky is Mizar or Zeta Ursæ Majoris, the second star in the 'tail' of the Great Bear – almost overhead as seen from Britain on February evenings. It is of the second magnitude, and close beside it is the fourth-magnitude Alcor (80 Ursæ Majoris). The separation is 708 seconds of arc, so that both members of the pair are very easy to see with the naked eye. Telescopically, Mizar itself is seen to be double, with one component decidedly brighter than the other. In fact Mizar was the first double to be found telescopically, by the seventeenth-century astronomer Joannes Riccioli (best remembered today for giving names to the craters of the Moon).

Strangely, the Arabs of a thousand years ago said that Alcor was difficult to see with the naked eye! Yet the Arabs were nothing if not keen-sighted; light pollution was much less of a problem in those times than it is now, and at present anyone with average sight can see Alcor with no difficulty at all. It does not seem likely that there has been any real alteration, so can there be another explanation?

Between the Mizar pair and Alcor is an 8th-magnitude star, first noted in 1681 by Georg Eimmart of Nürnberg. His observation caused no particular interest, and the next report came in 1723 from another German whose name has not been preserved. He believed it to be a new star or even a planet (!) and named it Sidus Ludovicianum, in honour of Ludwig V, Landgrave of Hesse. Can this, not Alcor, be the 'difficult' object of the Arabs?

Again it seems unlikely; Ludwig's Star is not connected with the Mizar group, as it lies well in the background, and it has shown no sign of variability. So the mystery remains – and today it is very clear that Sidus Ludovicianum is well beyond the range of naked-eye visibility.

E. A. MILNE. Our first *Yearbook* centenary of 1996 is that of Edward Arthur Milne, who was born in Hull on February 14, 1896. He graduated from Trinity College, Cambridge, and during the First World War carried out important theoretical work in anti-aircraft techniques (his eyesight was not good enough for military service). He then returned to Cambridge, and in 1925 became Professor of Applied Mathematics in Manchester. His last move was made in 1929, to Oxford, where he remained for the rest of his life, and accepted the Rouse Ball Chair of Mathematics there.

His main contributions were in cosmology and astrophysics. In 1932 he put forward his theory of Kinematic Relativity, which caused a great deal of interest and led to useful advances, though it is no longer regarded as a rival to Einstein's General Theory of Relativity. Milne died in Dublin, where he had been attending a conference, on September 21, 1950.

MARCH

New Moon: March 19 *Full Moon:* March 5

Equinox: March 20

Summer Time in Great Britain and Northern Ireland commences on March 31.

MERCURY is still well placed for observation as a morning object for the first half of the month, for observers in equatorial and southern latitudes, and they should refer to Figure 2 given with the notes for February. By the end of its period of visibility Mercury has increased in brightness to magnitude −0.8. For observers in northern temperate latitudes the planet remains unsuitably placed for observation.

VENUS, magnitude −4.3, continues to be visible as a brilliant evening object in the west-north-western sky.

MARS is in conjunction with the Sun on March 4 and is therefore unsuitably placed for observation.

JUPITER, magnitude −2.1, continues to be visible as a morning object in the south-eastern sky, for several hours before sunrise. Its path amongst the stars during 1996 is shown in Figure 6, given with the notes for June.

SATURN is too close to the Sun for observation as it passes through conjunction on March 17.

OPEN CLUSTERS IN CANCER. Cancer, the Crab, is one of the least prominent of the Zodiacal constellations, but it is easy to find, as it lies within the large triangle formed by Procyon, Regulus, and Pollux. In shape it looks a little like a very dim and ghostly Orion. Mythologically it represents a sea-crab which Juno, queen of Olympus, sent to the rescue of the multi-headed Hydra which was

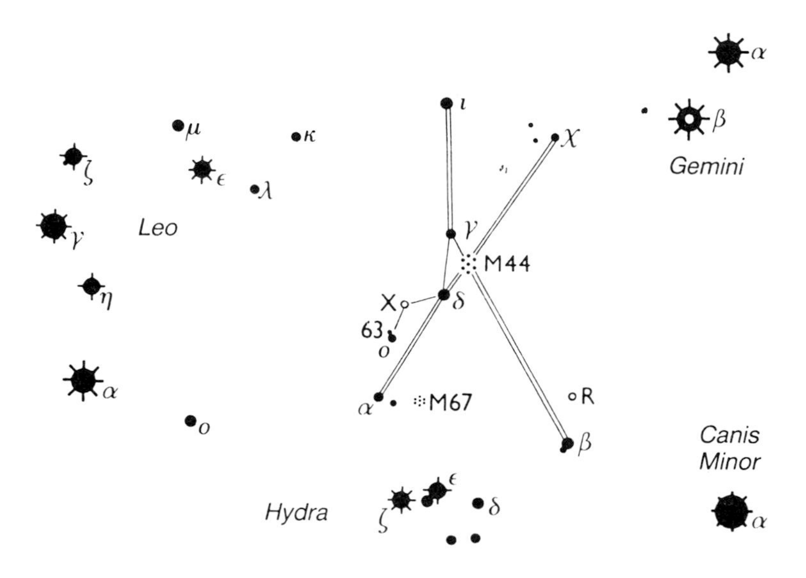

Figure 3. M.44 and M.67, the two clusters in Cancer.

being attacked by Hercules. Not unnaturally, Hercules trod on the crab and squashed it, but as a reward for its efforts Juno placed it in the sky (next to Leo, the Nemæan lion, another of Hercules' victims!).

Cancer has only two stars above the fourth magnitude, Beta (3.5) and Delta (3.9); next come Iota (4.0), Alpha or Acubens (4.2), and Gamma (4.7). However, the constellation is redeemed by the presence of two splendid open clusters, M.44 (Præsepe) and M.67.

Præsepe is an easy naked-eye object, and has been known since ancient times; Aratus (250–300 BC) wrote that rain threatened if Præsepe looked dim in an apparently clear sky. Hipparchus (around 130 BC) called it a 'cloudy star', and it was also catalogued by Ptolemy (around AD 150), but it was only with the telescopic work of Galileo in 1610 that its true nature became clear. Galileo recorded that it 'contains not one star only, but a mass of more than forty small stars'. In fact, the total number is much greater than this. Messier saw it in 1769 as 'a cluster of stars'.

Præsepe is often nicknamed the Beehive. Another name for it is the Manger, and the stars which flank it – Delta and Gamma Cancri

– are the Aselli or 'Asses'. Præsepe is about 520 light-years away, and is devoid of nebulosity, so that it is obviously older than the Pleiades cluster; the true diameter is over 15 light-years. As it is one and a half degrees in apparent diameter, the best views of it are probably to be obtained with binoculars or a very low-power eyepiece on a telescope.

The other cluster is M.67, which is in the same 7 × 50 binocular field with Acubens. It is on the fringe of naked-eye visibility, and is very easy in binoculars; it is not hard to resolve into stars. Its distance is well over 2500 light-years, and it is, in fact, a much larger and more populous cluster than Præsepe; it was discovered by Köhler around 1775, and in 1780 was described by Messier as 'a cluster of small stars'.

The main interest of M.67 is its age. Open clusters do not persist for very long on the cosmical scale, as they are disrupted by the gravitational effects of non-cluster stars, but M.67 lies at least 1400 light-years above the galactic plane, so that there are fewer 'field stars' to disrupt it. It is almost certainly the oldest open cluster known.

Another object in the Crab worth finding is X Cancri, which is in the same 7 × 50 binocular field with Delta. It is a semi-regular variable with a range of from magnitude 5.5 to 7.5 and a rough period of about 195 days. Its spectral type is N, and it is one of the very reddest of all stars, so that with binoculars it may be identified at once.

APRIL

New Moon: April 17 *Full Moon:* April 4

MERCURY, having passed through superior conjunction on March 28, is now moving eastwards from the Sun and it becomes possible to observe it in the early evening sky. For observers in northern temperate latitudes this will be the most favourable evening apparition of the year. Figure 4 shows, for observers in latitude N.52°, the changes in azimuth (true bearing from north through east, south, and west) and altitude of Mercury on successive evenings when the Sun is 6° below the horizon. This condition is known as the end of evening civil twilight and in this latitude and at this time of year occurs about 35 minutes after sunset. The changes in the brightness of the planet are indicated by the relative sizes of the circles marking Mercury's position at five-day intervals. It will be noticed that Mercury is at its brightest before it reaches greatest eastern elongation (20°) on April 23.

VENUS continues to be visible as a brilliant evening object, magnitude −4.4, completely dominating the western sky for several hours after sunset. It reaches its greatest eastern elongation (46°) on April 1. For observers in the British Isles Venus will be setting after 23^h and towards the end of the month observers in Scotland will have the unusual chance of seeing Venus just above the north-western horizon after midnight!

MARS is still too close to the Sun for observation.

JUPITER continues to be visible as a morning object in the south-eastern sky. By the end of the month observers in the latitudes of the British Isles should be able to see it just above the horizon, by 02^h. Jupiter's magnitude is −2.3.

SATURN, magnitude +1.1, becomes visible as a difficult morning object at the beginning of the month, being low above the eastern horizon before twilight inhibits observation. However, observers in

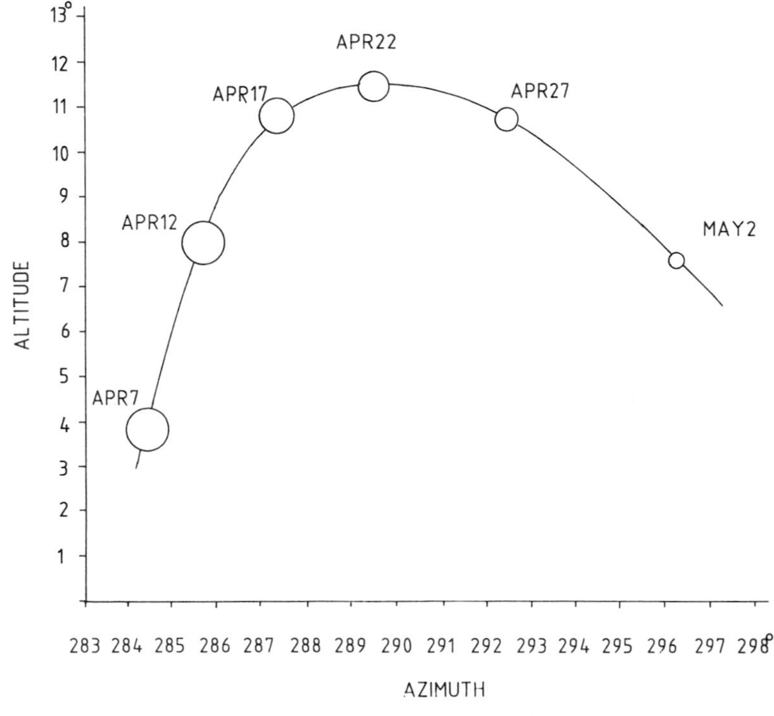

Figure 4. Evening apparition of Mercury for latitude N.52°.

the British Isles will have to wait until early in May before they can glimpse the planet.

Thatcher's Comet and the Lyrids. The main April meteor shower is that of the Lyrids. Maximum activity occurs on April 20–21, and the usual Zenithal Hourly Rate, or ZHR, is about 10, though there have been exceptional displays in the past – notably in 1922 and 1982, and, apparently, earlier in 1803. The shower is of fairly brief duration, and lasts for much less than a week. The position of the radiant is RA 18^h 8^m, dec. $\pm 32°$. (ZHR is a measure of meteor activity, and represents the number of naked-eye shower meteors which would be expected to be seen by an observer under ideal conditions, with the radiant at the zenith. As these conditions are never fulfilled, the actual observed rate is always less than the theoretical ZHR.)

There seems no doubt that the parent comet of the Lyrids is Thatcher's Comet, 1861 I. It was discovered on April 5, 1861 by the American observer A. E. Thatcher (no connection with any modern politician) when it was of magnitude 7.5. It slowly brightened up, and by May had reached magnitude 3.5, with a tail at least a degree long. Between May 9 and 10 it rose almost to the second magnitude, and passed within 50,000,000 kilometres of the Earth. During early June it was lost in the twilight, but was recovered from the Southern Hemisphere in July, and was followed until September 7, when the magnitude had dropped to 10. An orbit was worked out, and the period established as 415 years, so that we have no hope of seeing the comet again – even though we can still follow its debris in the form of the Lyrids seen every April.

APRIL CENTENARIES. The Russian astronomer Nicholas Theodor Bobrovnikoff was born at Starobielsk on April 29, 1896, and educated at Kharkov University, subsequently emigrating to America and in 1934 becoming Professor at the Wesleyan University, Delaware. He specialized in cometary studies, and in 1929 published an important paper in which he worked out the rate at which periodical comets fade with time. He found that they cannot persist for more than a million years, which meant that if they had been formed at the same time as the planets, and had kept in their present orbits, they would have dissipated long ago (today it is generally thought that the short-period comets come from the Kuiper Cloud, not so very far beyond the orbit of Neptune, and the longer-period comets from the much more remote Oort Cloud, at least a light-year from the Sun). Bobrovnikoff died in 1988.

Our second centenary is that of Carl Nicholas Adalbert Krüger, who was born at Marienberg (Germany) in 1832 and died on April 21, 1896. He was educated in Berlin, and in 1853 became assistant to Argelander at Bonn; he went to Helsinki in 1862 and finally, in 1876, to Gotha. He determined a number of stellar parallaxes, and in 1893 published a useful catalogue of 2153 red stars.

TWO LUNAR ECLIPSES. This month's eclipse is the first of two total lunar eclipses this year (the second is in September). The last time that this happened was in 1993 (June 4 and November 29); the next will be in 2000 (January 21 and July 16). There will be no umbral lunar eclipses in 1998.

VENUS AT ELONGATION. Venus reaches elongation on April 1. Theoretically it should then be at half-phase, but because of the well-known Schröter effect dichotomy at eastern elongations is always early; probably it will be reached during the last week of March. The effect is due to Venus' dense, extensive atmosphere.

MAY

New Moon: May 17 *Full Moon:* May 3

MERCURY is moving towards inferior conjunction on May 15 but observers in northern temperate latitudes may be able to detect the planet low above the west-north-western horizon at the end of evening civil twilight, though only for the first few days of the month. They should refer to the diagram given with the notes for April. During the last week of the month Mercury becomes visible in the eastern sky in the mornings before dawn, though only for observers in equatorial and southern latitudes. These observers should be careful not to confuse Mercury with Mars which is also emerging slowly from the morning twilight (see notes for next month).

VENUS is still a brilliant evening object in the western sky in the evenings, magnitude −4.5, though the period of time available for observation is decreasing, particularly for observers in northern temperate latitudes.

MARS is still too close to the Sun for most of May but during the last week of the month it should be possible for observers in equatorial latitudes to see it low in the eastern sky in the early mornings before the increasing twilight inhibits observation. Its magnitude is +1.3.

JUPITER, magnitude −2.5, continues to be a conspicuous object in the south-eastern sky, in the mornings.

SATURN, magnitude +1.0, already in view to those further south, gradually becomes visible to observers in the latitudes of the British Isles during May, when they may detect it low above the east-south-east horizon before the morning twilight inhibits observation. Saturn is in Pisces.

PLUTO AT OPPOSITION. Pluto comes to opposition on May 22. It is only 7° south of the celestial equator, and so is well placed, but the magnitude is only about 14, so that a telescope of reasonable size is needed to show it.

Pluto has puzzled astronomers ever since Clyde Tombaugh identified it, from the Lowell Observatory, in 1930. Its diameter is a mere 2324 kilometres, so that it is smaller than our Moon and also smaller than the senior satellite of Neptune, Triton; it is accompanied by a companion, Charon, which is 1270 kilometres across. Since Charon's diameter is more than half that of Pluto, it can hardly be classed as a normal satellite, and its orbital period of 6.3 days is the same as Pluto's axial rotation period – so that to a Plutonian observer, Charon would remain 'fixed' in the sky. Even when combined, Pluto and Charon could not possibly cause measurable perturbations in the movements of giants such as Uranus and Neptune – yet it was on the basis of such perturbations that Percival Lowell made a reasonably accurate prediction of Pluto's position. Either this was sheer luck, or else the real 'Planet X' remains to be found.

But is Pluto a true planet? In size, it is not; and it has a strange orbit, carrying it between 4425 million and 7375 million kilometres from the Sun in a period of 247.7 years. The orbital inclination is 17°, so that although its orbit brings it within that of Neptune there is no fear of a collision. Between 1979 and 1999 Neptune, not Pluto, is 'the outermost planet'; Pluto passed perihelion in 1989.

In recent years several smaller bodies, with diameters of the order of 150 to 200 kilometres, have been found in these remote parts of the Solar System. They are thought to come from the Kuiper Belt, and there is increasing support for the idea that Pluto is itself a Kuiper Belt object. The same may be true of Triton, which was presumably captured by Neptune in the remote past while Pluto remains independent.

Pluto has a very tenuous but surprisingly extensive atmosphere; Charon apparently has none. Pluto contributes 80 per cent of the total light of the system; the maximum separation between it and Charon is only 0″.8. Unfortunately no probes to them have been funded as yet.

CORVUS AND CRATER. Ptolemy, last of the great astronomers of Classical times, listed 48 constellations, all of which are still to be found on our maps though in many cases the boundaries have been altered. Some of Ptolemy's groups are surprisingly obscure. Corvus, the Crow, is admittedly fairly easy to find, because its four

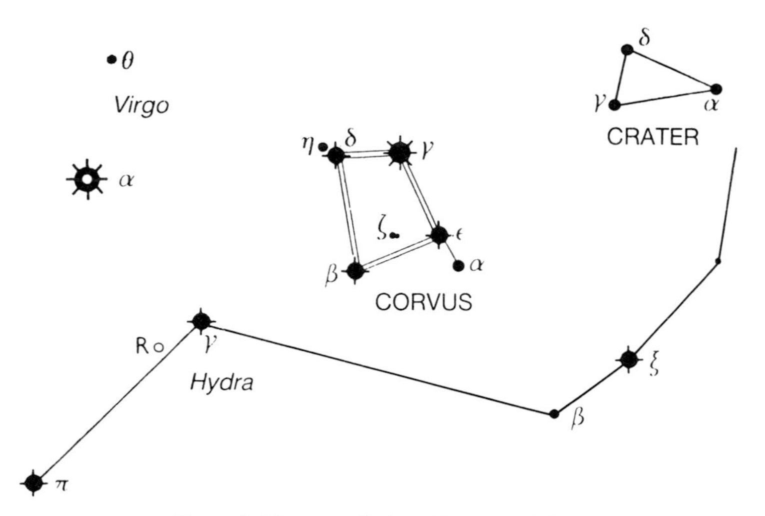

Figure 5. The constellations Corvus and Crater.

leading stars (Gamma, Beta, Delta, and Epsilon) are arranged in a quadrangle, and are between magnitudes 2½ and 3 (the star lettered Alpha is only of magnitude 4). Adjoining it is Crater, the Cup, with only one star (Delta, 3.6) above the fourth magnitude. Both these groups lie on Hydra, the Watersnake, which is actually the largest constellation in the entire sky.

It must be admitted that both Corvus and Crater are remarkably devoid of interesting objects! All four leaders of Corvus have been suspected of being slightly variable, but this has not been confirmed. Epsilon Corvi has a K-type spectrum, and binoculars show it to be decidedly orange.

JUNE

New Moon: June 16 *Full Moon:* June 1

Solstice: June 21

MERCURY is at greatest western elongation (24°) on June 10 and is therefore a morning object until the very end of the month, though only for observers in the equatorial regions and the Southern Hemisphere. These observers should be able to detect Mercury above the east-north-east horizon about half-an-hour before sunrise. There is a noticeable increase in Mercury's brightness during the month, the magnitude increasing from +1.6 to −1.1. Mercury and Mars are only a few degrees apart for most of the month but for the second half of June they are both outshone by Venus as it moves rapidly outwards from the Sun. The table below gives the magnitudes and elongations from the Sun.

Date	*Mercury*		*Venus*		*Mars*	
	Elong.	Mag.	Elong.	Mag.	Elong.	Mag.
May 27	W17	+2.4	E22	−4.3	W18	+1.4
June 6	23	+1.0	E7	−3.9	20	+1.4
16	23	+0.1	W8	−3.9	23	+1.4
26	W17	−0.7	W22	−4.3	W25	+1.4

To complicate matters Mercury passes 3°N. of Aldebaran on June 21, and Mars passes 6°N. of it on June 27!

VENUS, magnitude around −4.0, is now moving rapidly towards the Sun and will be seen in the evenings only for a short while, low in the western evening sky after sunset, for the first few days of the month. It passes rapidly through inferior conjunction on June 10. On this occasion Venus passes 30′ south of the centre of the Sun. At the next two June inferior conjunctions (in 2004 and 2012) the distance is so small as to give rise to transits of Venus across the disk of the Sun. For the second half of the month Venus may be seen as a

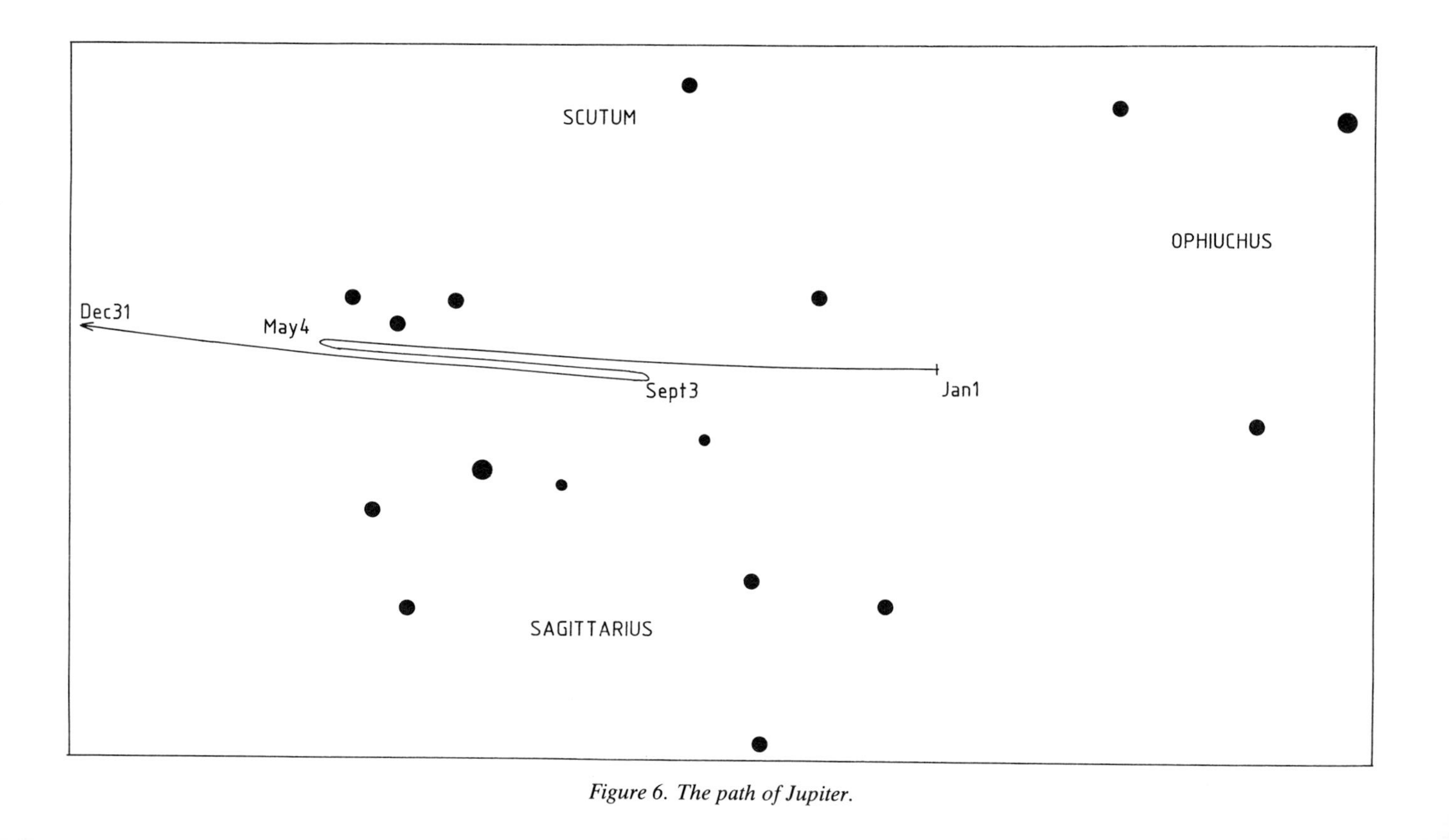

Figure 6. The path of Jupiter.

morning object, magnitude −4.2, low above the eastern horizon before dawn.

MARS, magnitude +1.4, is a difficult morning object, low in the eastern sky before the increasing twilight renders observation impossible. Mars passes between the Pleiades and Hyades during the second half of the month. Observers in northern temperate latitudes will not be able to see the planet until next month.

JUPITER is a conspicuous object in the night sky, magnitude −2.7.

SATURN, magnitude +0.9, continues to be visible as a morning object in the south-eastern sky before dawn.

THE GREAT COMET COLLISION. In July 1994 Comet Shoemaker–Levy 9 impacted Jupiter, causing a tremendous amount of interest among the general public as well as among astronomers. Nothing of the kind had ever been seen before, though no doubt similar collisions have happened often enough in the past.

The full analyses of the observations will take a long time to complete, and will be reported in a future *Yearbook*. Meantime, some preliminary conclusions can be drawn. The main point, of course, is that the effects on Jupiter, striking though they were, were purely temporary. At present (spring 1995) the results of the impact can still be seen, but are decreasing, and will certainly disappear before very long. After all, the fragments of the cometary nucleus were only a few kilometres across at most, whereas the diameter of Jupiter is about ten times that of the Earth.

It is true that if a similar collision had happened to the Earth, there would have been widespread devastation – and it has often been suggested that it was an impact, some 65,000,000 years ago, which caused a climatic change great enough to wipe out the dinosaurs. But Jupiter is a much larger target than the Earth, and the danger of a Shoemaker–Levy-type impact on our world in the foreseeable future is very slight indeed.

Jupiter is well placed for southern observers this month, though it is very low down as seen from Britain and the main United States. Astronomers, both professional and amateur, will be keeping a close watch on it to see just what happens there. During the next few years the planet will move northwards again; the opposition declinations will be as follows:

1996, −23: 1997, −17: 1998, −4: 1999, +10: 2000, +20: 2002, +23

Jupiter's synodic period is 399 days, and usually there is an opposition each year – but 2001 is an exception, as oppositions fall on November 28, 2000 and January 1, 2002. The last year without an opposition of Jupiter was 1990.

THE ZODIAC. In 1995 there was a curious report that astronomers had discovered a new constellation of the Zodiac! It is, of course, true that Ophiuchus (the Serpent-bearer) does cross the Zodiac between Scorpius and Sagittarius, but there is nothing new in this and, in fact, some details were given in the April notes for the *1995 Yearbook*. Remember, too, that the constellation patterns are entirely arbitrary. If we had followed, say, the Egyptian pattern we would have had a Cat and a Hippopotamus, though the stars would have been exactly the same. Only astrologers pay any attention to the Zodiacal signs – and astrology can only be described as rubbish.

Ophiuchus crosses the Equator as well as the Zodiac. Of its leading stars, Alpha (Rasalgethi) has a declination of +12°33′, while Theta Ophiuchi, magnitude 3.3, is at declination −25°, only about one degree further north than Antares, the red supergiant leader of the Scorpion.

TRANSITS OF VENUS. The last transit of Venus occurred in 1882. It therefore seems fairly safe to say that nobody now living can remember it. If memory goes back to the age of three, then such an observer would by now have reached the advanced age of 117 – and though it is true that there are reports of human beings older than this, it is hardly likely that any of them would have been looking at the Sun in 1882!

JULY

New Moon: July 15 *Full Moon:* July 1, 30

EARTH is at aphelion (furthest from the Sun) on July 5, at a distance of 152 million kilometres.

MERCURY is at superior conjunction on July 11 and is too close to the Sun for observation for observers in the Northern Hemisphere. However, for those near to and south of the Equator the planet will be visible in the evening sky during the last week of the month and reference should be made to the diagram given with the notes for August.

VENUS, magnitude −4.5, is visible as a morning object in the eastern sky before sunrise, rapidly pulling away from the Sun. By the end of the month Venus is rising about two hours before the Sun. Venus continues to move away from the Sun at a faster rate than Mars and by the end of July they are about 10° apart.

MARS continues to be visible as a morning object, magnitude +1.5, in the eastern sky before dawn. It is readily available for observation by those in equatorial and southern latitudes but becomes visible to those in northern temperate latitudes only very gradually during the month. The path of Mars amongst the stars is shown in the diagram given with the notes for November.

JUPITER, magnitude −2.7, is now at its brightest since it is at opposition on July 4 and therefore available for observation throughout the night. It is a disappointing object for observers in northern temperate latitudes because of its low altitude even when on the meridian (for example, the maximum altitude as seen from southern England is only 16°, the same as the midwinter Sun). When closest to the Earth its distance is 626 million kilometres. The path of Jupiter among the stars during its 1996 apparition is shown in Figure 6, given with the notes for June.

SATURN continues to be visible as a morning object, magnitude +0.8. By the middle of the month observers in the British Isles should be able to see it low in the eastern sky before midnight.

URANUS is at opposition on July 25, in Capricornus. The planet is only just visible to the naked-eye under the best of conditions since its magnitude is +5.7. In a small telescope it appears as a slightly greenish disk. At opposition Uranus is 2806 million kilometres from the Earth.

NEPTUNE is at opposition on July 18, in Sagittarius. It is not visible to the naked eye since its magnitude is +7.9. At opposition Neptune is 4360 million kilometres from the Earth.

THE NON-IDENTICAL TWINS. Both Uranus and Neptune come to opposition this month, and they are not far apart in the sky. In size and mass they are near-twins, but they are not identical. Uranus is essentially bland, Neptune dynamic.

Details are as follows:

	Uranus	*Neptune*
Equatorial diameter, km	51,118	50,538
Polar diameter, km	49,946	49,600
Mass, Earth = 1	14.5	17.2
Volume, Earth = 1	67	57
Reciprocal mass, Sun = 1	22,800	19,300
Escape velocity, km/s	22.5	23.9
Surface gravity, Earth = 1	1.17	1.2

Therefore Uranus is slightly the larger, but Neptune is the more massive. Yet they are so alike that one would have expected them to be similar in all respects. This is not so. In particular, Neptune – like Jupiter and Saturn – has a strong inner heat source, while Uranus has not. Voyager images (and later images from the Hubble Space Telescope) show that the greenish disk of Uranus is remarkably lacking in detail, while Voyager showed a huge dark spot on Neptune together with other features. In fact Neptune's Great Dark Spot was the same size, relative to the planet, as the Great Red Spot is to Jupiter. Recent Hubble images have not shown the

Great Dark Spot, and we cannot be sure that it still exists, but before long we may hope to find out.

Another major difference is the axial inclination – about 29° to the perpendicular for Neptune, but 98° for Uranus, so that Uranus' rotation is technically retrograde. Just why Uranus has this exceptional tilt is not known. The favoured theory is that in the remote past it suffered a gigantic impact and was literally toppled over. This does not sound plausible – but it is hard to think of any alternative.

The magnetic axes are strongly inclined to the rotational axes, and do not pass through the centres of the globes, so that in this respect the two really are alike; both have magnetic fields, and both have rings, though Neptune's are very ghostly. The satellite systems are completely different. Uranus has a grand total of fifteen, of which only four (Ariel, Umbriel, Titania, and Oberon) are over 1000 km in diameter; Neptune has only one large satellite, Triton (diameter 2705 km) and seven small ones. It seems almost certain that Triton is not a genuine satellite at all, but a Kuiper Belt object which was captured, while one of the small attendants, Nereid, has a highly eccentric orbit more like that of a comet than a satellite.

Only one probe, Voyager 2, has surveyed these two giants from close range, and when another mission will be sent there remains to be seen. Let us hope that it will not be too long delayed, because we do not yet have anything like a full knowledge of these lonely worlds on the outpost of the Solar System.

THE CELESTIAL ARCHER. Sagittarius, the Archer, has no first-magnitude star, but it is very rich, and the glorious star-clouds hide our view of the mysterious centre of the Galaxy. Unfortunately Sagittarius is always very low from Britain, and part of it never rises at all. July evenings give Britons their best chance; but from countries such as Australia and New Zealand Sagittarius is almost overhead, and can be seen in its full glory. It has no distinctive pattern, though it is often said to give the impression of the shape of a teapot!

AUGUST

New Moon: August 14 *Full Moon:* August 28

MERCURY is moving eastwards, away from the Sun: reaching its greatest eastern elongation (27°) on August 21. Since it was at aphelion only two days earlier it must be at almost its maximum possible angular distance from the Sun. Actually it is also moving southwards in declination so that it will not be suitably placed for observation in northern temperate latitudes. For observers in southern latitudes this will be the most favourable evening apparition of the year. Figure 7 shows, for observers in latitude S.35°, the changes in azimuth (true bearing from the north through east, south, and west) and altitude of Mercury on successive evenings when the Sun is 6° below the horizon. This condition is known as the end of evening civil twilight and in this latitude and at this time of year occurs about 30 minutes after sunset. The changes in the brightness of the planet are indicated by the relative sizes of the circles marking Mercury's position at five-day intervals. It will be noticed that Mercury is at its brightest before it reaches greatest eastern elongation. On the first morning of the month Mercury is very close to Regulus, the star being about 1½ magnitudes fainter than Mercury and only a few tenths of a degree to the left, as seen from South Africa. From Australasia, where dawn occurs 6–10 hours earlier, Regulus will be several tenths of a degree higher in altitude, with much less difference in azimuth.

VENUS, magnitude −4.3, is a magnificent object in the eastern sky in the mornings before dawn, reaching its greatest western elongation (46°) on August 20. However, Mars continues to move farther out from the Sun and by the end of August the two planets are only a few degrees apart.

MARS, magnitude +1.5, continues to be visible as a morning object in the eastern sky before dawn. Mars is moving eastwards in Gemini and by the very end of the month is passing about 6°S. of Pollux.

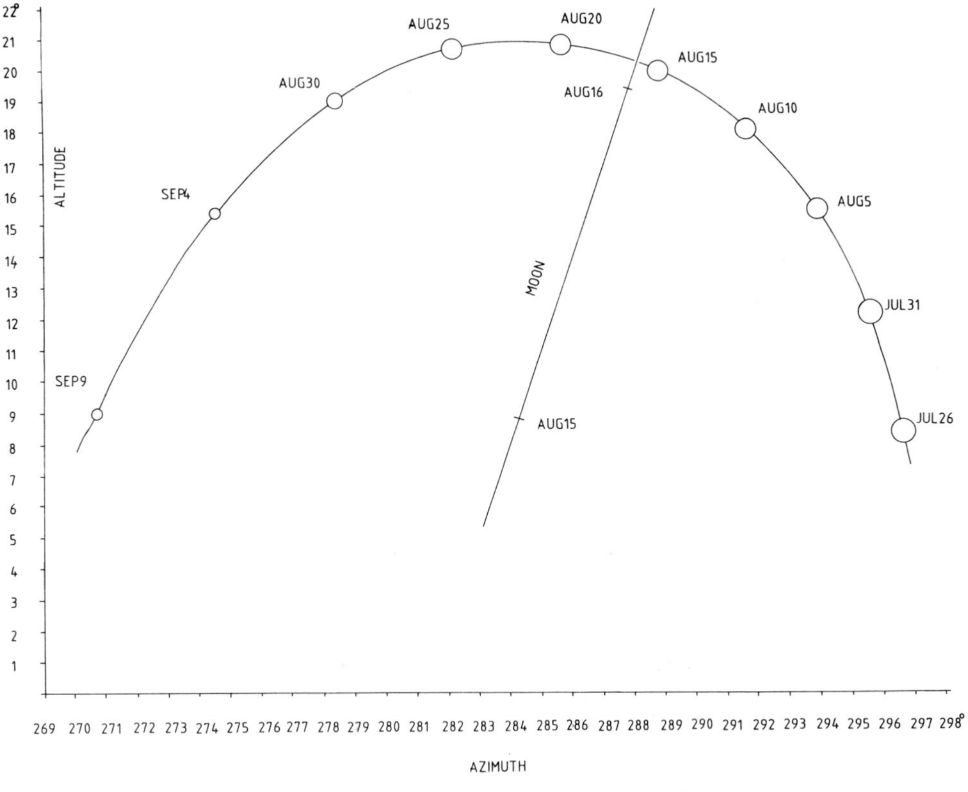

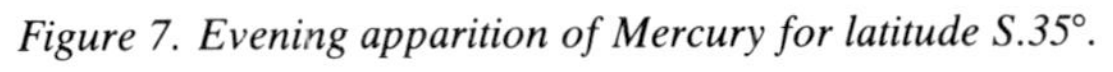

Figure 7. Evening apparition of Mercury for latitude S.35°.

JUPITER is a conspicuous object in the southern skies in the evenings, magnitude −2.6.

SATURN, magnitude +0.7, is now visible for the greater part of the night as it approaches opposition towards the end of next month. Saturn is on the borders of the constellations of Cetus and Pisces.

MERCURY – ACCORDING TO ANTONIADI. The one space-probe which has surveyed Mercury from close range, Mariner 10 (1974–5) has told us that the planet is totally barren, with a maximum surface temperature of over 600 degrees C., and an atmosphere which is virtually non-existent; the pressure is probably about 10^{-10} millibar, and the atmosphere is technically an exosphere. Craters are plentiful, and there is one huge basin, the Caloris Basin, 1300 km in diameter. But pre-Mariner, the best map of Mercury was provided by the Greek astronomer Eugenios Antoniadi, who spent much of his life in France and observed with the great Meudon refractor. Antoniadi's book about Mercury was published in 1934, though it was not until 1974 that it appeared in English (in fact I undertook the translation). Antoniadi's views about the planet sound strange today. He wrote:

> 'The dawn and dusk so peculiar to Mercury would seem excessively pale. . . . A more-than-glacial breeze would be extremely punishing during the night, while a wind incomparably more scorching than the desert simoom would probably give rise to a spectacle of fuming dunes, which during the day would raise eddies of greyish dust which would cover the sky, and conceal even the Sun with sinister, all-absorbing clouds. . . . The constellations will, of course, present the same aspect from Mercury as from the Earth; but they will show up more brilliantly, and the Milky Way will appear so low in altitude that it would seem to be nearly on top of the observer. Star-twinkling will be unknown. . . . The Zodiacal Light, in which Mercury is always plunged, will seem to be more luminous but more diffuse than it does to us; and since it will fill most of the sky, it will form a very broad band, somewhat ill-defined. . . . Seen from Mercury, all the planets are naturally superior, coming to opposition with the Sun. Venus will be the most brilliant star in the sky, much more brilliant than it appears from Earth – a veritable celestial diamond, casting shadows with diffraction fringes. Its disk at

opposition will attain a diameter of 70″ when Mercury is near aphelion. Our Earth will come next, a magnificent star of the first magnitude, whose maximum diameter will attain 33″, and which, with its bright surface, should be comparable with our own view of Venus, casting comparable shadows. Keen eyesight would also show our Moon, slowly oscillating from one side of the Earth to the other. Mars, as seen from Mercury, would never fall to the 2nd magnitude; Jupiter and Saturn would be of the first magnitude for most of the time. . . . Mercury would be an ideal planet from which to admire and study comets. Indeed, these "hairy stars" would appear, near their perihelion, with a grandeur which would make our comets of 1811, 1843, 1858, and 1882 seem very feeble. . . . Meteor showers ought to be rich from Mercury, but will, in general, describe very rapid trajectories in the rarefied Mercurian atmosphere.'

In fact, we now know that Mercury's atmosphere is too thin to produce luminous meteors, and there are no winds. Antoniadi's description does at least show us how little we knew about the innermost planet only little more than half a century ago!

THE ROSSE REFLECTOR. It is hoped that around August 1996 the great Rosse 72-inch reflector, at Birr Castle, will be almost ready to use once more. Built in 1845 by the third Earl of Rosse, this was the telescope used to discover the spiral nature of the objects we now know to be galaxies. In 1909 it was taken out of use, but restoration has been in progress, and if all goes well the 'Leviathan' will again be turned skyward at some time during the second half of 1996.

THE PERSEIDS. Do not forget the August Perseids, much the most reliable of all annual meteor showers. Moonlight will not interfere this year, so we may confidently expect a good display of cosmic fireworks.

SEPTEMBER

New Moon: September 12 *Full Moon:* September 27

Equinox: September 22

MERCURY, for observers in the Southern Hemisphere, remains visible as an early evening object for the first ten days of the month, and they should refer to the diagram given with the notes for August. Mercury passes through inferior conjunction on September 17, moving rapidly westwards and becoming visible in the morning sky before dawn for the last week of the month. Observers should refer to the diagram given with the notes for October.

VENUS, magnitude −4.2, continues to be visible as a brilliant object in the eastern morning sky before dawn. Mars continues to increase its elongation from the Sun and on September 3 is only 3°N. of Venus. The two planets are over 10° apart by the end of the month.

MARS is still a morning object, magnitude +1.5, visible in the eastern sky for several hours before sunrise.

JUPITER, magnitude −2.4, continues to be visible as a conspicuously bright object in the south-western skies in the evenings. For observers in the latitudes of the British Isles it is lost to view well before midnight.

SATURN, magnitude +0.5, reaches opposition on September 26 and, as a result, is visible throughout the hours of darkness. After the Earth's passage through the ring plane last year the rings are now beginning to open up. The minor axis of the rings is now 4 arcseconds; the polar diameter of Saturn is 17 arcseconds. At opposition Saturn is 1271 million kilometres from the Earth. The diagram (Figure 8) shows the path of Saturn amongst the stars during the year.

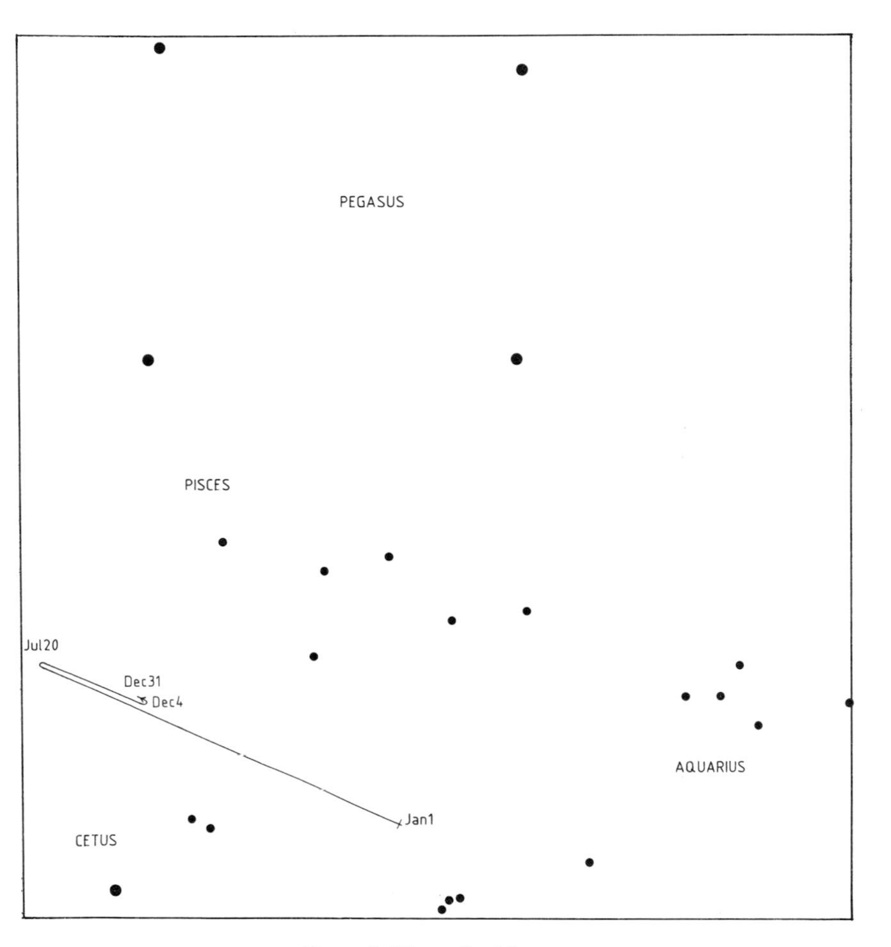

Figure 8. The path of Saturn.

LINNÉ AND THE LUNAR ECLIPSE. Lunar eclipses are neither so important nor so spectacular as total eclipses of the Sun, but they are well worth watching, and the colours can often be beautiful. Obviously, the aspect of the eclipse depends entirely upon conditions in the Earth's atmosphere through which all sunlight reaching the eclipsed Moon has to pass.

The main effect on the Moon itself is that a wave of cold sweeps across the lunar surface; the temperature plunges dramatically. In the past it has been suggested that this has a perceptible effect upon some of the formations, notably Linné in the Mare Serenitatis, the well-formed 'sea' adjoining the Lunar Apennines.

One American astronomer of the early twentieth century, W. H. Pickering, believed that Linné increased in visibility and that some sort of icy coating was responsible. Linné had been described as a deep craterlet by the first two great selenographers, Beer and

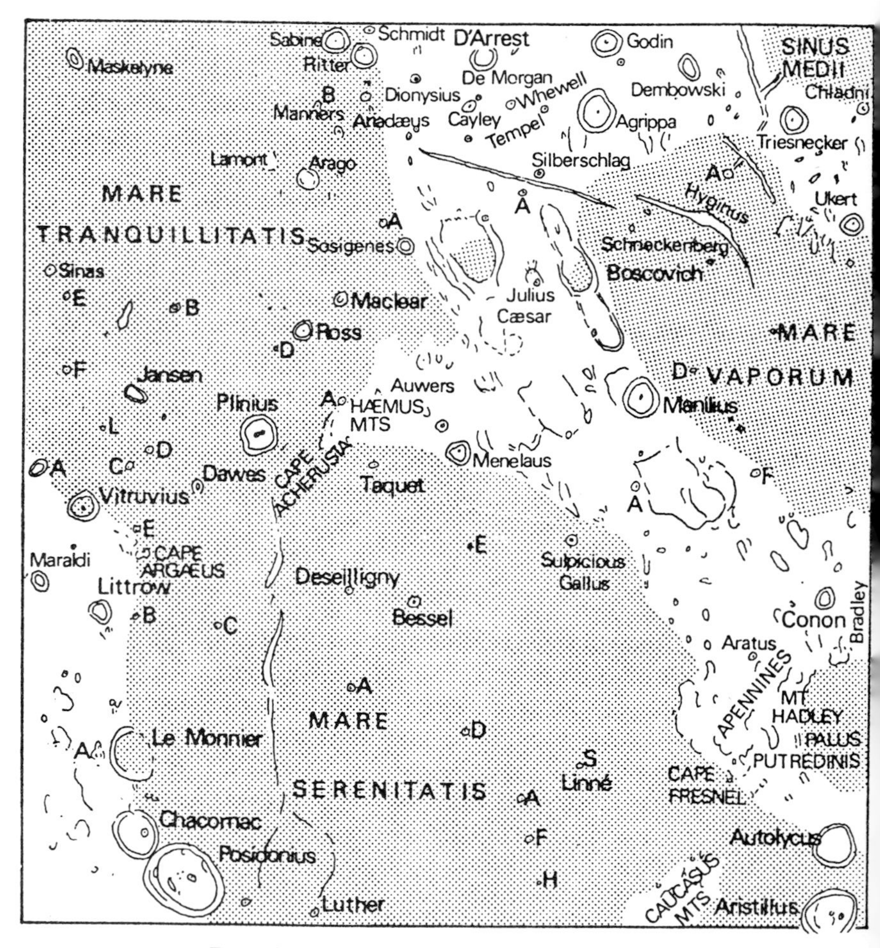

Figure 9. Section of lunar map showing Linné (S).

Mädler, who published their major book in 1838; in 1866 Julius Schmidt reported that Linné had changed into a white spot, and it was this which led to a revival of interest in lunar observation. Linné has been shown by the space-probes to be a normal bowl-shaped crater surrounded by a whitish patch, and it seems certain that no change has occurred there; but Pickering was convinced that eclipses had a marked effect, and lunar observers still tend to look at Linné before, during, and after an eclipse, though without any real hope of detecting anything unusual.

ARMAND FIZEAU. Armand Hippolyte Louis Fizeau was born in Paris in 1819, and co-operated with Foucault in securing the first good photographs of the Sun in 1845. He re-determined the velocity of light, and in 1848 developed Doppler's principle, showing that spectral lines shift according to the velocity of the light-source: to the long-wave or red end if the source is receding, to the blue or short-wave end if the source is approaching. Christian Doppler had described this effect for sound-waves (everyone is familiar with the demonstration of a passing ambulance or fire-engine sounding its horn!) but it was Fizeau who applied it to light, and from this point of view it would be fairer to refer to it as the Doppler–Fizeau effect.

OCTOBER

New Moon: October 12 *Full Moon:* October 26

Summer Time in Great Britain and Northern Ireland ends on October 27.

MERCURY reaches greatest western elongation (18°) on October 3 and continues to be visible low above the eastern horizon before dawn. For observers in northern temperate latitudes this will be the most favourable morning apparition of the year. Figure 10 shows, for observers in latitude N.52°, the changes in azimuth (true bearing from the north through east, south, and west) and altitude of Mercury on successive mornings when the Sun is 6° below the horizon. This condition is known as the beginning of morning civil twilight and in this latitude and at this time of year occurs about 35 minutes before sunrise. The changes in the brightness of the planet are indicated by the relative sizes of the circles marking Mercury's position at five-day intervals. It will be noticed that Mercury is at its brightest after it reaches western elongation. After the middle of the month it becomes too close to the Sun to be observed as it moves towards superior conjunction early next month.

VENUS is still a brilliant object in the eastern morning sky before dawn. Its magnitude is −4.1.

MARS, magnitude +1.4, continues to be visible as a morning object. It moves steadily eastwards from Cancer into Leo, passing only 0°.2 south of Regulus on October 4.

JUPITER is visible in the south-western sky in the early part of the evening. Its magnitude is −2.2.

SATURN, magnitude +0.6, is just past opposition and continues to be visible as a noticeable object in the southern skies for the greater part of the night.

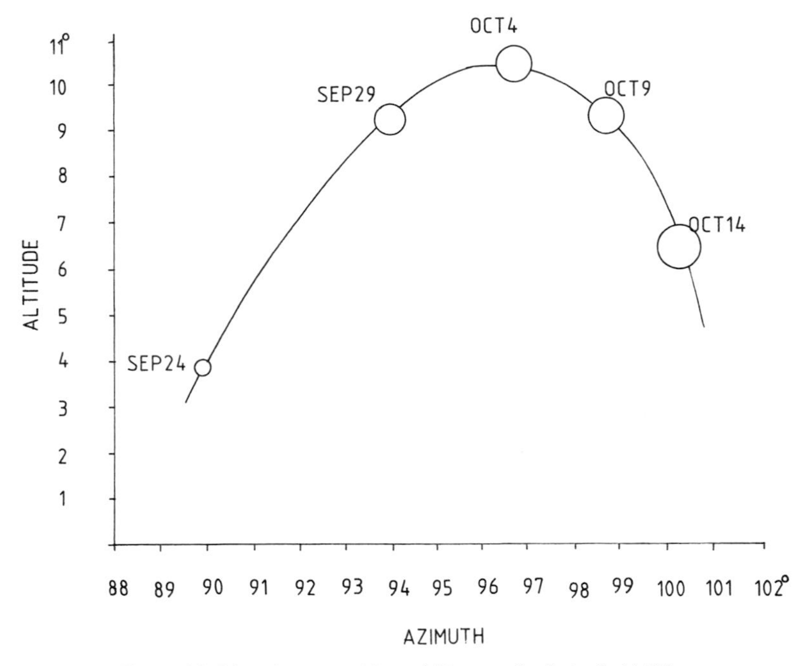

Figure 10. Morning apparition of Mercury for latitude N.52°.

THE TRIANGULUM SPIRAL. This is the best time of the year for looking at the Great Spiral in Andromeda, M.31. From Britain, it is high in the south after sunset; it attains a reasonable altitude from most of Australia and South Africa, and only from southern New Zealand is it to all intents and purposes lost. But how many people also pause to look at the nearby spiral, M.33 in Triangulum?

Triangulum itself is easy to find, between Almaak (Gamma Andromedæ) and Hamal (Alpha Arietis); its three main stars really do form a triangle – Alpha (magnitude 3.4), Beta (3.0), and Gamma (4.0). Gamma makes a nice little pair with its neighbour Delta (4.9). But Triangulum is notable mainly because of the presence of M.33, which was discovered by Messier himself in 1764 and described by him as a nebula of 'whitish light of almost even brightness. However, along ⅔ of its diameter it is a little brighter. Contains no star. Seen with difficulty in a 1-foot telescope.' Lord Rosse, with his 72-inch reflector, called it 'full of knots. Spiral arrangement. Two similar curves like an S cross in the centre.'

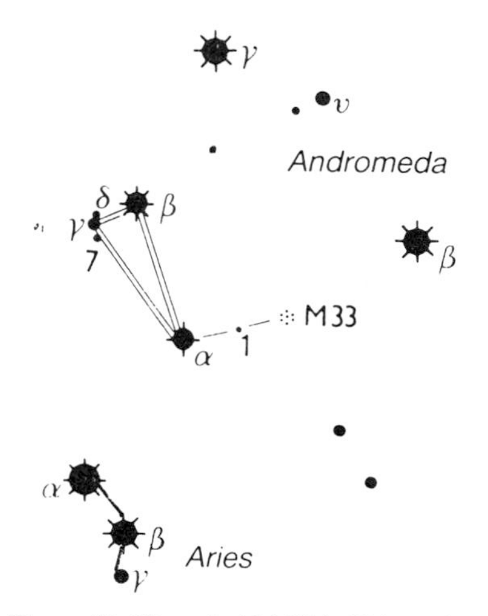

Figure 11. The spiral M.33 in Triangulum.

M.33 is a member of the Local Group, and is 2,350,000 light-years away, slightly further than M.31. It is also smaller than M.31, and much smaller than our Galaxy, but it is almost face-on to us, whereas M.31 lies at a narrow angle and the full beauty of the spiral is lost. M.33 is of type Sc, so that it is comparatively loose.

It has been claimed that M.33 can be seen with the naked eye (on the next clear night, try to glimpse it!). With binoculars it is easy to locate. With 7 × 50 binoculars it is in the same field as Alpha; look past the fainter star 1 Trianguli, and at about an equal distance beyond it, in the direction of Beta Andromedæ, you will see M.33 as a fairly large but decidedly dim haze. Oddly enough, this low surface brightness makes it rather elusive telescopically, and a good method is to locate it with binoculars and then use the telescope.

Mars in Leo. Early in October Mars is close to Regulus, in Leo, and is about the same brightness, though of course, the colours are very different; Mars is red, Regulus pure white. The conjunction is a good chance for astrophotographers to take a series of interesting pictures, showing how Mars moves relative to the star.

Centenaries. Two anniversaries are worthy of note this month. Didrik Magnus Axel Möller, a leading Swedish astronomer, was born at Schonen in 1830, and was educated at Lund University, where he remained throughout his career. He was appointed Professor of Astronomy in 1863, and carried out valuable mathematical work, mainly in connection with planetary perturbations. He died at Lund a hundred years ago, on October 25, 1896.

François Félix Tisserand was born at Naivty-sur-Georges (France) in 1845, and in 1866 joined the staff of the Paris Observatory. He became Director of the Toulouse Observatory in 1870, and returned to Paris as Director in 1892. He, too, was primarily a mathematician, but he also made skilful observations of the planets, particularly Mars. He died on October 20, 1896.

THE START OF THE SPACE AGE. How many people now recall that the Space Age began on October 4, 1957, with the launch of Sputnik 1, Russia's first artificial satellite? We have come a long way since then, and it is sobering to remember that even after the end of the war there were still eminent scientists who were utterly convinced that space research would never be possible – just as in 1902 a leading American astronomer, Simon Newcomb, proved to his complete satisfaction that no heavier-than-air machine could ever fly!

NOVEMBER

New Moon: November 11 *Full Moon:* November 25

MERCURY passes through superior conjunction on November 2 and is therefore unobservable at first. It is not until the last week of the month that it can be seen in the evenings – and then only by observers in equatorial and southern latitudes. Then they should be able to detect Mercury low above the west-south-west horizon about half-an-hour after sunset, magnitude about −0.5.

VENUS, magnitude −4.0, continues to be visible as a brilliant object in the eastern sky before sunrise.

MARS remains a morning object in Leo, magnitude +1.1. Its slight, reddish tinge is a useful aid to identification. Figure 12 shows the path of Mars amongst the stars during the second half of the year.

JUPITER, magnitude −2.0, remains visible in the south-western sky in the early part of the evening. Observers with a good pair of binoculars, providing that they are steadily supported, should attempt to detect the four Galilean satellites. The main difficulty in observing them is the overpowering brightness of Jupiter itself.

SATURN is an evening object, magnitude +0.8. Its path amongst the stars is shown in Figure 8, given with the notes for September.

ZETA PHŒNICIS. To northern observers the prototype eclipsing binary, Algol, is well placed this month; its normal magnitude is 2.1, but every 2½ days it dips down to below the third magnitude as its less luminous component partially eclipses the brighter star. Not many Algol stars exceed the fifth magnitude at maximum – in fact there are only four:

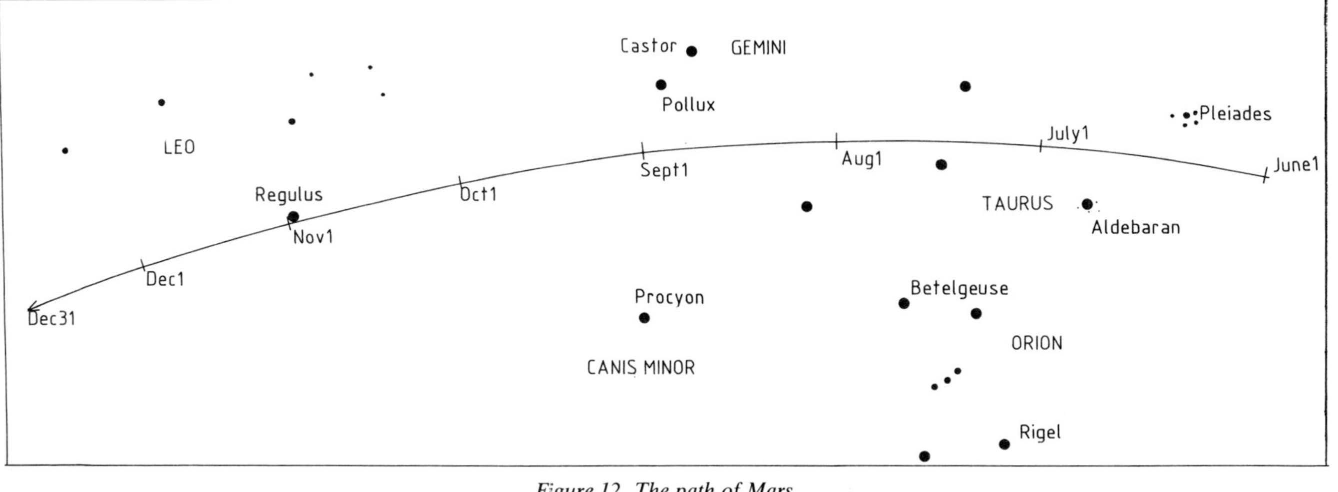

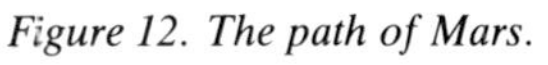
Figure 12. The path of Mars.

Star	*Range*	*Period, days*	*Spectrum*
Algol (Beta Persei)	2.1–3.4	2.9	B + G
Lambda Tauri	3.3–3.8	3.9	B + A
Zeta Phœnicis	3.9–4.4	1.7	B + B
Delta Libræ	4.9–5.9	2.3	B

Phœnix, the Phœnix, is one of the Southern Birds. Its brightest star, Alpha Phœnicis (Ankaa) is of magnitude 2.4. The constellation is not very distinctive, but there are two fairly obvious triangles, and the group lies between Achernar and Fomalhaut; as the declination of Ankaa is −42°, it is not visible from Britain at any time. Zeta, the Algol variable, lies near the brilliant Achernar. Its range is less than that of Algol, but its period is shorter, and the fluctuations are easy to follow; suitable comparison stars are Eta (4.4), Delta (3.9), and Chi Eridani (3.7).

In the same binocular field as Iota (4.7) lies the rapid pulsating variable SX Phœnicis, which has a period of only 79 minutes – one of

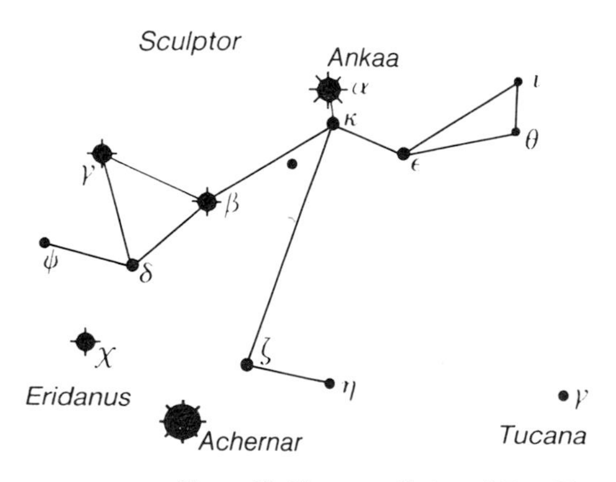

Figure 13. The constellation of Zeta Phœnicis.

the shortest known; its range is between magnitudes 7.1 and 7.5. It seems to be an A-type sub-dwarf, no more than three times as luminous as the Sun, lying at a distance of about 140 light-years. It has an annual proper motion of nearly 0″.9, and it was indeed this large proper motion which first brought it to the attention of astronomers; its rapid fluctuations were discovered in 1952 by O. G.

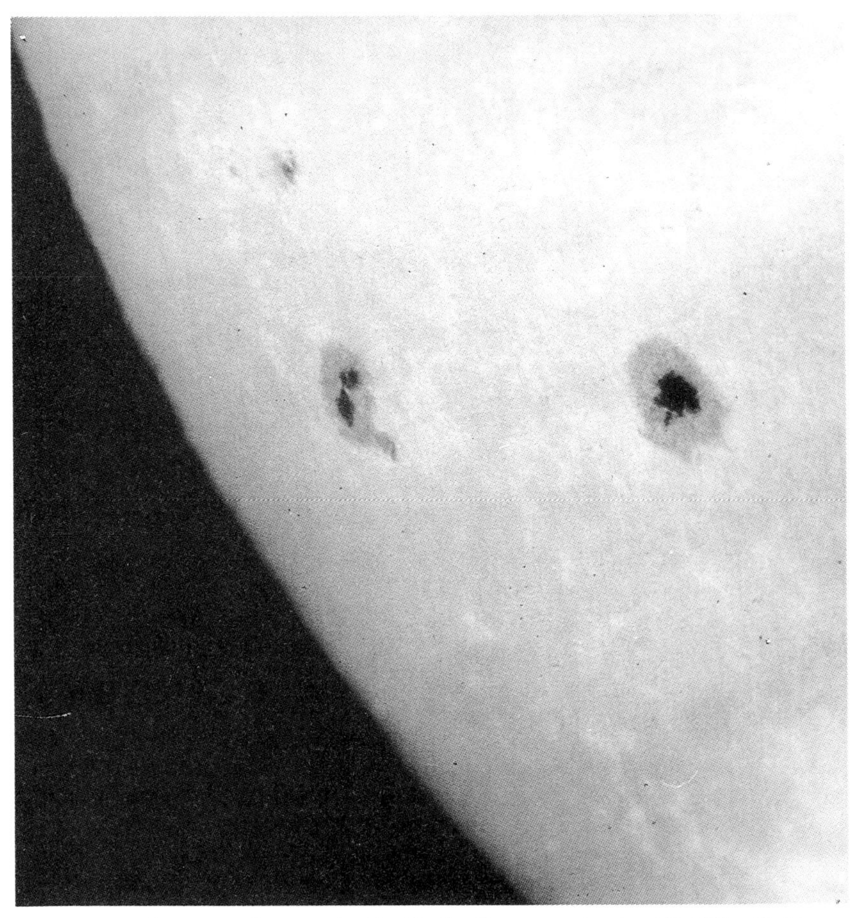

Figure 14. Sunspots, photographed by W. M. Baxter on October 17, 1959 with his 4-inch refractor.

Eggen, from the Canberra Observatory. It has a rather low mass, about 0.2 to 0.3 that of the Sun.

W. M. BAXTER. November 30 is the centenary of the birth of William Morley Baxter, one of the best-known of all amateur observers of the Sun. He was born at Brighton in Sussex, and trained as an electrical and mechanical engineer; during the First World War he served in the R.N.A.S. Subsequently he went to India, and after several years of valuable experience he returned to London to join the branch of the firm for which he had worked. However, it was after his retirement that he became well-known in astronomical circles. He had married in 1928, and with his wife settled in Acton in north London, where he set up an observatory, equipping it with a fine 5-inch refractor. He specialized in solar work, and his sunspot pictures are to be found in observatories all over the world. He became Director of the Solar Section of the British Astronomical Association in 1964, and at various times served as the Association's secretary and vice-president. In 1967 he was awarded the Association's Merlin Medal. He wrote one book, *The Sun and the Amateur Astronomer*, which became something of a classic. He continued his solar work until a few months before his death on December 9, 1971.

HUGO GYLDÉN. Gyldén, a leading Finnish astronomer, was born in Helsinki in 1841. He acted as assistant first at Helsinki and then at Pulkovo Observatory before going on to Sweden as Director of the Stockholm Observatory. He was an expert mathematician, and in 1871 came to the conclusion that the Galaxy is rotating – in which he was, of course, quite correct. He died on November 9, 1896.

DECEMBER

New Moon: December 10 *Full Moon:* December 24

Solstice: December 21

MERCURY reaches its greatest eastern elongation (20°) on December 15, but it is so far south of the Equator that it is not suitably placed for observation by those in northern temperate latitudes. However, observers in equatorial and southern latitudes will be able to see Mercury in the west-south-western sky after sunset for all except the last few days of the month; its magnitude decreasing from −0.5 to +1.5.

VENUS continues to be visible as a brilliant object, magnitude −4.0, in the eastern sky in the mornings before sunrise though the period available for observation is shortening noticeably as it moves through the Sun. By the end of the year it is not likely to be visible until after 07^h, for observers in southern England, while in the north of Scotland it will be nearly an hour later.

MARS, magnitude +0.8, continues to be visible as a morning object.

JUPITER, magnitude −1.9, is approaching the end of its 1996 apparition and only visible for a short while in the early evenings low above the south-western horizon. Because of its southern declination observers in northern temperate latitudes will only be able to see it for the first half of the month.

SATURN, magnitude +1.0, continues to be visible in the south-western sky in the evenings. By the middle of the month it is too low above the western horizon to be visible after midnight.

HUGH PERCIVAL WILKINS. Percy Wilkins (as he was always known) was born on December 4, 1896 at Carmarthen, and was educated at Carmarthen Grammar School. He became an engineer; after serving in the Army during the First World War he settled in Kent, and worked at the Ministry of Defence. On retirement he devoted all his time to studying the Moon, and produced a 300-inch lunar map together with a large book containing a full description of the surface features. He also wrote several popular books, and for many years was the Director of the Lunar Section of the British Astronomical Association. He died on January 23, 1960.

BRIGHT M-TYPE STARS. Orion is back, with its red supergiant Betelgeux. Betelgeux is of type M, and its colour is very evident. It is interesting to compile a list of other M-type stars above the fourth magnitude:

Star	*Magnitude*	*Type*	*Luminosity* (Sun = 1)	*Distance* (lt-years)	
β Andromedæ	2.1	M0	115	88	Mirach
λ Aquarii	3.7	M2	130	230	
σ Canis Majoris	3.5	M0	15,000	7500	
R Carinæ	var.	M5	var.	800	
μ Cephei	var.	M2	52,500	1560	
α Ceti	2.5	M2	130	130	Menkar
o Ceti	var.	Md	var.	95	Mira
γ Crucis	1.6	M3	160	88	
λ Draconis	3.8	M0	110	210	
τ Eridani	3.7	M3	130	235	
η Geminorum	var.	M3	130	186	Propus
μ Geminorum	2.9	M3	130	230	Teja'
β Gruis	2.1	M3	170	750	Al Dhanab
α Herculis	var.	M5	700	218	Rasalgethi
R Hydræ	var.	M1	var.	160	
γ Hydri	3.2	M0	120	160	
σ Libræ	3.3	M4	132	166	Zubenalgubi
α Lyncis	3.1	M0	120	170	
δ Ophiuchi	2.7	M1	130	140	Yed Prior
α Orionis	var.	M2	15,000	310	Betelgeux
β Pegasi	var.	M2	300	179	Scheat
ρ Persei	var.	M4	130	196	Gorgones Terti
L^2 Puppis	var.	M5	1500	75	
δ Sagittæ	3.8	M2	700	550	
η Sagittarii	3.1	M3	750	420	
α Scorpii	1.0	M3	7500	330	Antares
μ Ursæ Majoris	3.1	M0	120	156	Tania Australis
δ Virginis	3.5	M3	130	147	Minelauva

The luminosities and distances are taken from the Cambridge catalogue. Most of these stars are variable, and of course Mira,

R Hydræ, and R Carinæ fall well below naked-eye visibility at minimum. In many cases the colours are not evident with the naked eye, but binoculars bring them out well, and a survey of M-type stars is well worth undertaking.

Eclipses in 1996

There will be four eclipses in 1996, two of the Sun and two of the Moon.

1. *A total eclipse of the Moon on April 3–4* is visible from the extreme western part of Australia, Indonesia, Asia, Indian Ocean, Africa, the Atlantic Ocean, Europe (including the British Isles), Iceland, Greenland, Antarctica, the Americas (except the western part of North America), and the eastern Pacific. The eclipse begins at $3^{d}\ 22^{h}\ 21^{m}$ and ends at $4^{d}\ 01^{h}\ 59^{m}$. Totality lasts from $3^{d}\ 23^{h}\ 26^{m}$ to $4^{d}\ 00^{h}\ 53^{m}$.

2. *A partial eclipse of the Sun on April 17–18* is visible from New Zealand (except the extreme northern part), the South Pacific Ocean, and a small part of Antarctica. The eclipse begins at $17^{d}\ 20^{h}\ 31^{m}$ and ends at $18^{d}\ 00^{h}\ 43^{m}$. At maximum eclipse 0.88 of the Sun is obscured.

3. *A total eclipse of the Moon on September 27* is visible from western Asia, the western Indian Ocean, Africa, Antarctica, Europe (including the British Isles), Iceland, Greenland, the Atlantic Ocean, the Americas, and the Pacific Ocean (except the western part). The eclipse begins at $1^{h}\ 12^{m}$ and ends at $4^{h}\ 36^{m}$. The total phase begins at $2^{h}\ 19^{m}$ and ends at $3^{h}\ 29^{m}$.

4. *A partial eclipse of the Sun on October 12* is visible from north-east Canada, Greenland, Iceland, Europe, extreme western Asia, and North Africa. The eclipse begins at $11^{h}\ 59^{m}$ and ends at $16^{h}\ 05^{m}$. From Greenwich the eclipse begins at $12^{h}\ 59^{m}$ and ends at $15^{h}\ 31^{m}$, while from Edinburgh the eclipse begins and ends 10 minutes earlier than at Greenwich. Observers in the British Isles will see that about 60 per cent of the Sun is obscured at maximum.

Occultations in 1996

In the course of its journey round the sky each month, the Moon passes in front of all the stars in its path, and the timing of these occultations is useful in fixing the position and motion of the Moon. The Moon's orbit is tilted at more than five degrees to the ecliptic, but it is not fixed in space. It twists steadily westwards at a rate of about twenty degrees a year, a complete revolution taking 18.6 years, during which time all the stars that lie within about six and a half degrees of the ecliptic will be occulted. The occultations of any one star continue month after month until the Moon's path has twisted away from the star, but only a few of these occultations will be visible at any one place in hours of darkness.

There are five occultations of bright planets in 1996, two of Mercury, two of Venus and one of Mars.

Only four first-magnitude stars are near enough to the ecliptic to be occulted by the Moon: these are Regulus, Aldebaran, Spica and Antares. Only Aldebaran undergoes an occultation (six times) in 1996.

Predictions of these occultations are made on a world-wide basis for all stars down to magnitude 7.5, and sometimes even fainter. The British Astronomical Association has just produced the first complete lunar occultation prediction package for microcomputer users.

Recently occultations of stars by planets (including minor planets) and satellites have aroused considerable attention.

The exact timing of such events gives valuable information about positions, sizes, orbits, atmospheres and sometimes of the presence of satellites. The discovery of the rings of Uranus in 1977 was the unexpected result of the observations made of a predicted occultation of a faint star by Uranus. The duration of an occultation by a satellite or minor planet is quite small (usually of the order of a minute or less). If observations are made from a number of stations it is possible to deduce the size of the planet.

The observations need to be made either photoelectrically or visually. The high accuracy of the method can readily be appreciated when one realizes that even a stop-watch timing accurate to $0^s.1$ is, on average, equivalent to an accuracy of about 1 kilometre in the chord measured across the minor planet.

Comets in 1996

The appearance of a bright comet is a rare event which can never be predicted in advance, because this class of object travels round the Sun in an enormous orbit with a period which may well be many thousands of years. There are therefore no records of the previous appearances of these bodies, and we are unable to follow their wanderings through space.

Comets of short period, on the other hand, return at regular intervals, and attract a good deal of attention from astronomers. Unfortunately they are all faint objects, and are recovered and followed by photographic methods using large telescopes. Most of these short-period comets travel in orbits of small inclination which reach out to the orbit of Jupiter, and it is this planet which is mainly responsible for the severe perturbations which many of these comets undergo. Unlike the planets, comets may be seen in any part of the sky, but since their distances from the Earth are similar to those of the planets their apparent movements in the sky are also somewhat similar, and some of them may be followed for long periods of time.

The following periodic comets are expected to return to perihelion in 1995, and to be brighter than magnitude +15.

Comet	*Year of discovery*	*Period (years)*	*Predicted date of perihelion 1996*
Churyumov-Gerasimenko	1969	6.6	Jan. 17
Kopff	1906	6.4	July 2
Gunn	1970	6.8	July 24
Shoemaker-Holt (2)	1988	8.0	Aug. 20
Wild (4)	1990	6.2	Aug. 31
Machholz (1)	1986	5.2	Oct. 15
IRAS	1983	13.3	Oct. 31

Minor Planets in 1996

Although many thousands of minor planets (asteroids) are known to exist, only 3,000 of these have well-determined orbits and are listed in the catalogues. Most of these orbits lie entirely between the orbits of Mars and Jupiter. All of these bodies are quite small, and even the largest, Ceres, is only about 960 kilometres in diameter. Thus, they are necessarily faint objects, and although a number of them are within the reach of a small telescope few of them ever reach any considerable brightness. The first four that were discovered are named Ceres, Pallas, Juno and Vesta. Actually the largest four minor planets are Ceres, Pallas, Vesta and Hygeia (excluding 2060 Chiron, which orbits mainly between the paths of Saturn and Uranus, and whose nature is uncertain). Vesta can occasionally be seen with the naked eye and this is most likely to occur when an opposition occurs near June, since Vesta would then be at perihelion. Ephemerides for these minor planets in 1996 are:

1 Ceres

1996		2000.0 Right Ascension		Declin-ation		Geocentric distance	Visual mag.	Elong-ation
month	*day*	*hr.*	*min.*	°	′			°
2	17	16	28.55	−15	42.6	2.688	8.7	79.8W
2	27	16	38.91	−16	7.9	2.561	8.6	87.3W
3	8	16	47.61	−16	28.4	2.434	8.5	95.2W
3	18	16	54.39	−16	45.1	2.308	8.4	103.4W
3	28	16	58.93	−16	59.2	2.187	8.2	112.2W
4	7	17	.99	−17	11.8	2.073	8.1	121.5W
4	17	17	.38	−17	23.9	1.972	7.9	131.4W
4	27	16	57.03	−17	36.4	1.886	7.7	141.8W
5	7	16	51.14	−17	49.6	1.820	7.5	152.8W
5	17	16	43.17	−18	3.6	1.778	7.3	164.2W
5	27	16	33.88	−18	18.4	1.762	7.0	175.0W
6	6	16	24.29	−18	34.4	1.773	7.1	171.2E
6	16	16	15.40	−18	52.1	1.812	7.4	160.0E
6	26	16	8.13	−19	12.5	1.875	7.6	148.9E
7	6	16	3.09	−19	36.3	1.961	7.9	138.3E
7	16	16	.55	−20	4.1	2.065	8.1	128.3E
7	26	16	.62	−20	36.0	2.184	8.3	118.9E
8	5	16	3.16	−21	11.3	2.313	8.4	110.0E
8	15	16	7.98	−21	49.5	2.450	8.6	101.7E
8	25	16	14.88	−22	29.3	2.590	8.7	93.7E

2 Pallas 2000.0

1996		*Right Ascension*		*Declin-ation*		*Geocentric distance*	*Visual mag.*	*Elong-ation*
month	*day*	*hr.*	*min.*	°	′			°
2	17	14	46.60	+ 3	14.6	1.988	8.7	108.5W
2	27	14	52.00	+ 5	51.7	1.900	8.6	116.9W
3	8	14	54.77	+ 8	48.1	1.827	8.5	125.3W
3	18	14	54.76	+11	57.2	1.771	8.3	133.3W
3	28	14	51.99	+15	8.8	1.737	8.2	140.0W
4	7	14	46.79	+18	10.3	1.728	8.2	144.5W
4	17	14	39.76	+20	49.0	1.743	8.2	145.5W
4	27	14	31.77	+22	54.4	1.783	8.3	143.1W
5	7	14	23.84	+24	20.6	1.846	8.5	138.0E
5	17	14	16.87	+25	7.0	1.929	8.6	131.5E
5	27	14	11.57	+25	16.8	2.028	8.8	124.4E
6	6	14	8.37	+24	56.0	2.140	9.0	117.2E
6	16	14	7.40	+24	10.9	2.263	9.2	110.0E
6	26	14	8.62	+23	7.4	2.392	9.3	103.2E
7	6	14	11.88	+21	50.9	2.525	9.5	96.5E
7	16	14	16.95	+20	25.4	2.661	9.6	90.2E
7	26	14	23.62	+18	54.4	2.796	9.7	84.1E
8	5	14	31.69	+17	20.6	2.930	9.8	78.2E

3 Juno 2000.0

1996		*Right Ascension*		*Declin-ation*		*Geocentric distance*	*Visual mag.*	*Elong-ation*
month	*day*	*hr.*	*min.*	°	′			°
7	27	0	54.06	+ 5	53.7	1.724	9.2	109.6W
8	6	1	1.78	+ 5	39.6	1.594	9.0	117.5W
8	16	1	7.44	+ 5	2.5	1.473	8.8	126.0W
8	26	1	10.70	+ 3	59.9	1.365	8.5	135.2W
9	5	1	11.36	+ 2	31.2	1.272	8.2	145.1W
9	15	1	9.33	+ 0	38.5	1.198	7.9	155.5W
9	25	1	4.93	− 1	31.4	1.146	7.7	165.6W
10	5	0	58.90	− 3	47.2	1.118	7.5	170.7W
10	15	0	52.31	− 5	54.4	1.116	7.6	163.8E
10	25	0	46.48	− 7	39.3	1.137	7.8	153.2E
11	4	0	42.51	− 8	52.6	1.180	8.0	142.5E
11	14	0	41.16	− 9	30.7	1.240	8.2	132.4E
11	24	0	42.80	− 9	34.8	1.314	8.4	123.0E
12	4	0	47.40	− 9	8.8	1.398	8.5	114.4E
12	14	0	54.74	− 8	17.6	1.490	8.7	106.5E

4 Vesta 2000.0

1996		*Right Ascension*		*Declination*		*Geocentric distance*	*Visual mag.*	*Elongation*
month	*day*	*hr.*	*min.*	°	′			°
1	23	14	43.54	− 8	5.9	2.164	7.6	81.4W
2	2	14	57.41	− 8	43.2	2.036	7.4	88.0W
2	12	15	9.95	− 9	9.1	1.908	7.3	94.9W
2	22	15	20.84	− 9	23.3	1.783	7.1	102.3W
3	3	15	29.69	− 9	25.9	1.661	6.9	110.0W
3	13	15	36.11	− 9	17.6	1.547	6.7	118.3W
3	23	15	39.69	− 8	59.4	1.441	6.5	127.2W
4	2	15	40.12	− 8	33.2	1.349	6.3	136.7W
4	12	15	37.29	− 8	1.8	1.272	6.1	146.8W
4	22	15	31.37	− 7	29.4	1.214	5.8	157.0W
5	2	15	23.04	− 7	1.1	1.178	5.6	166.1W
5	12	15	13.43	− 6	42.4	1.166	5.6	168.5E
5	22	15	3.87	− 6	38.1	1.178	5.7	161.2E
6	1	14	55.74	− 6	51.0	1.213	5.9	151.2E
6	11	14	50.03	− 7	21.6	1.268	6.1	141.1E
6	21	14	47.30	− 8	8.6	1.340	6.3	131.5E
7	1	14	47.71	− 9	9.3	1.426	6.5	122.6E
7	11	14	51.13	−10	20.7	1.521	6.7	114.3E
7	21	14	57.31	−11	39.6	1.624	6.9	106.7E
7	31	15	5.97	−13	3.4	1.732	7.1	99.6E

A vigorous campaign for observing the occultations of stars by the minor planets has produced improved values for the dimensions of some of them, as well as the suggestion that some of these planets may be accompanied by satellites. Many of these observations have been made photoelectrically. However, amateur observers have found renewed interest in the minor planets since it has been shown that their visual timings of an occultation of a star by a minor planet are accurate enough to lead to reliable determinations of diameter. As a consequence many groups of observers all over the world are now organizing themselves for expeditions should the predicted track of such an occultation cross their country.

In 1984 the British Astronomical Association formed a special Asteroid and Remote Planets Section.

Meteors in 1996

Meteors ('shooting stars') may be seen on any clear moonless night, but on certain nights of the year their number increases noticeably. This occurs when the Earth chances to intersect a concentration of meteoric dust moving in an orbit around the Sun. If the dust is well spread out in space, the resulting shower of meteors may last for several days. The word 'shower' must not be misinterpreted – only on very rare occasions have the meteors been so numerous as to resemble snowflakes falling.

If the meteor tracks are marked on a star map and traced backwards, a number of them will be found to intersect in a point (or a small area of the sky) which marks the radiant of the shower. This gives the direction from which the meteors have come.

The following table gives some of the more easily observed showers with their radiants; interference by moonlight is shown by the letter M.

Limiting dates	*Shower*	*Maximum*	*R.A.*		*Dec.*	
			h	*m*	°	
Jan. 1–4	Quadrantids	Jan. 4	15	28	+50	M
April 20–22	Lyrids	Apr. 21	18	08	+32	
May 1–8	Aquarids	May 5	22	20	0	M
June 17–26	Ophiuchids	June 20	17	20	−20	
July 15–Aug. 15	Delta Aquarids	July 29	22	36	−17	M
July 15–Aug. 20	Piscis Australids	July 31	22	40	−30	M
July 15–Aug. 25	Capricornids	Aug. 2	20	36	−10	M
July 27–Aug. 17	Perseids	Aug. 12	3	04	+58	
Oct. 15–25	Orionids	Oct. 22	6	24	+15	
Oct. 26–Nov. 16	Taurids	Nov. 3	3	44	+14	
Nov. 15–19	Leonids	Nov. 17	10	08	+22	
Dec. 9–14	Geminids	Dec. 13	7	28	+32	
Dec. 17–24	Ursids	Dec. 23	14	28	+78	M

M = moonlight interferes

Some Events in 1997

ECLIPSES

There will be four eclipses, two of the Sun and two of the Moon.

March 8–9: total eclipse of the Sun – S.E. Asia, N.W. of N. America.
March 24: partial eclipse of the Moon – Africa, Europe, the Americas.
September 1–2: partial eclipse of the Sun – Australasia.
September 16: total eclipse of the Moon – Australasia, Asia, Africa, Europe.

THE PLANETS

Mercury may be seen more easily from northern latitudes in the evenings about the time of greatest eastern elongation (April 6) and in the mornings around greatest western elongation (September 16). In the Southern Hemisphere the corresponding most favourable dates are around January 24 (mornings) and August 4 (evenings).

Venus is visible in the mornings until the end of February and in the evenings from May to December.

Mars is at opposition on March 17.

Jupiter is at opposition on August 9.

Saturn is at opposition on October 10.

Uranus is at opposition on July 29.

Neptune is at opposition on July 21.

Pluto is at opposition on May 25.

The Universal Astronomer: David Allen, 1946–1994

FRED WATSON

Astronomers inhabit a curiously paradoxical world. Much of their time is spent in close confrontation with the infinite, encouraging a certain detachment from the comings and goings of life on this rocky left-over from the Sun's formation. Yet, inevitably, they are part and parcel of it.

You can probably imagine the kind of thing I mean. One minute, you are gazing with awe on the filigree tracery of gravitationally lensed arcs of light from the most distant galaxies, imaged by the Hubble Space Telescope. The next, you are staring with disbelief at the less-than-delicate scribblings of your offspring on the newly painted dining-room wall. One minute, you are lavishing infinite care on a wafer-thin glass photograph containing half a million star and galaxy images; the next, you are clumsily spilling coffee down your clothes. And, one minute, you are marvelling at the pinpoint accuracy with which orbital dynamics can predict a cosmic event like the impact of comet Shoemaker–Levy 9 with Jupiter, while the next, you are painfully aware that none of us *really* knows what lies just around the corner.

It was a confusion of thoughts like these that tumbled through my mind one sunny morning back in August 1993. I had been jolted back from somewhere near the galactic centre by a bulletin message on my computer screen in Cambridge. It was from David Allen, a friend and colleague at the Anglo-Australian Observatory, a brilliant and accomplished astronomer well known to regular readers of this *Yearbook*. It was, it said, to bring news of impending surgery he had to undergo. 'A large and fast-growing tumour has taken hold in my brain. At present, the major symptom is total blindness over most of the left field of vision in both eyes. Other symptoms are gradually appearing . . .'

The news came as a body-blow. Only thirteen months earlier, I had watched, helpless, while my own mother's life was brought to

Figure 1. Dr David Allen in 1993, when he was awarded Australia's Eureka Prize for the Promotion of Science (Photo: C. Bento, © Australian Museum).

an end by a similar tumour. I knew, in one angry, aching moment, what the eventual consequence of David's message was likely to be.

An heroic struggle

To those who knew him, it came as no surprise that David put up a valiant fight against the marauding cancer. No less unexpected was the detached curiosity with which he tried to understand the mortal conflict going on inside his own head – the stamp of a true researcher. Following surgery and radiotherapy, the tumour 'merely shrugged its shoulders and grew back six months later'. Even in this desperate situation, David's way with words did not desert him.

The last time I saw him, during a visit to Australia in the golden Southern-Hemisphere autumn of 1994, he explained how, each morning, he reviewed his faculties, and spent the day working to recover those he had lost during the night. Feebly, I joked that he probably still outstripped most of us in mental ability. Apparently, I was not the first – and, it was almost certainly true. His last words to me were 'I'm still fighting. But I'm afraid I can't guarantee to win'. He was almost apologetic. He clearly felt as much compassion for the loved ones and friends he knew he was leaving as we did for him.

David died on July 26, 1994, at the tragically young age of 47. The accolades had been pouring in from all over the globe as news of his illness spread. At a moving reception at the Anglo-Australian Observatory library on August 4, his family and colleagues, together with friends from all facets of his life, heard many, many tributes.

Months have now passed since those sad events. As might be expected, obituaries have appeared in the astronomical press, and Australia's ABC Radio has devoted an edition of its *Science Show* to David's life and work. With characteristic consideration – not to mention an innate enthusiasm for having the facts right – he left comprehensive autobiographical notes to help.

It remains to pay tribute to David in the *Yearbook of Astronomy*. It is a publication that was very close to his heart, and almost every edition over the last twenty-five or so years has contained one of his articles. I feel privileged that the task of celebrating such a gifted and prolific life has fallen to me, just one among his many friends. But how I wish that the need had never been there.

Born at the age of nine

David had good reason for claiming to have been 'born at the age of nine'. Though he entered the world on July 30, 1946, in the Manchester suburb of Altrincham, he spent seven years from the age of two with his body and legs encased in plaster as a result of a mis-diagnosed hip complaint. It's hard to imagine the effect on a child of such enforced paralysis. 'It made me tough' was David's simple conclusion, and tough he certainly was when it came to arguing his case or hiking over mountain tops – or, eventually, fighting for his life.

It was during those early, immobile years that the first spark of interest in astronomy flickered into life. An eclipse of the Moon impressed itself on his memory. And, in later years, he recalled with affection a set of coloured picture cards on astronomy that came free in packets of tea. I know about those: I collected them too, and, secure in the album that could be bought to hold them (price 6*d*), they still live in my book-case. It's impossible to look at them now without thinking of David.

At the age of sixteen, an operation to fuse his right hip left him with the slight limp that, alone, betrayed his troubled medical history. But it also gave him a new freedom to pursue the outdoor activities he cherished. Walking, particularly in mountain country, remained a life-long passion. Years later, he climbed Tanzania's Mount Kilimanjaro – a truly amazing feat – and he would doubtless have scaled even greater heights had time been on his side.

If David's self-effacing account is to be believed, his career at Manchester Grammar School was less than spectacular. It did, however, lead to an exhibition at Trinity Hall in the University of Cambridge – with the by-product of a telescope built in the school workshops during his final year. Astronomy was already a consuming passion, and it was while a schoolboy that he first corresponded with Patrick Moore – on the delicate subject of an error in one of the great man's moon-maps! That letter established David as a recognized lunar observer, and led to a life-long friendship between the two astronomers.

At Cambridge, David studied mathematics with physics, and gained an upper second-class honours degree. Perhaps more significantly, it was there he met his wife, Carol, whose support he always considered to be a pillar of his success. They clearly had a very special relationship, and the books they later wrote together were just one manifestation of that.

In October 1967, David moved to the University Observatories at Cambridge to begin research for his Ph.D. degree. His supervisor was Dr David Dewhirst, still at Cambridge today and still clearly proud of his former student, though he admitted to me that David needed little supervision – merely occasional guidance as to which of his abundant new ideas were the right ones to follow up. Dr Dewhirst recognized the exceptional talent of his student, but I doubt whether David himself perceived that he was standing on the brink of an exceptional career.

Infrared astronomy

It was soon after he began his Ph.D. studies that David wrote his first article for the *Yearbook of Astronomy*. It appeared in the 1969 edition, and it was entitled 'Infrared Astronomy'. The article was concerned with the invisible heat radiation beyond the red end of the rainbow of colours to which our eyes are sensitive. In the late 1960s, new detectors were only just opening up this infrared region to astronomers.

Red light has a wavelength of about 0.7 micrometres (0.7μm, or seven ten-thousandths of a millimetre), and infrared extends beyond this to the microwave region at about 1000μm (1 mm). But ground-based infrared observations are restricted to a number of atmospheric 'wavelength-windows' whose transparency is critically dependent on freedom from water vapour. It is this aspect that has driven infrared astronomers to the highest mountain sites to pursue their craft.

David's article dealt primarily with wavelengths beyond 8μm, the region we would now call the mid-infrared. In a remarkably accurate cameo preview of his subsequent career, it covered observations of virtually every class of celestial object.

His doctoral dissertation was submitted in July 1970, with the title *Infrared Studies of the Lunar Terrain*. A quarter of a century later, it still makes interesting reading. It reported infrared measurements of 'thermal anomalies' – areas of the Moon's surface that cool unusually slowly during the lunar night or an eclipse. David attributed the anomalies to boulder fields, and demonstrated that previous models of the lunar surface's thermal behaviour needed to be modified to account for the boulders scattered over it. 'May all your anomalies be little ones' was the benediction written by a friend at the outset of the work, and David was able to conclude in his dissertation that they were!

The observations for his Ph.D. research were carried out in the winters of 1968/69 and 1969/70 on the 0.75-metre telescope of the University of Minnesota's O'Brien Observatory. In later life, David recalled the extreme discomfort of those bitter winter nights in the unheated telescope dome, his body heat draining to the floor through the soles of his feet and anaesthetizing his lower limbs. But the rewards were great indeed. So little was known of the infrared sky that almost every observation resulted in an important discovery. David found himself the explorer in a new, uncharted country, and was transfixed by the vistas that opened up to him. More than anything else, these pioneering observations served to guide his career irrevocably into the brave new world of infrared astronomy.

To the Great South Land

After completing his Ph.D., David returned to the States on a Carnegie Fellowship to work with the famous Mount Wilson telescopes. Much of 1971 was spent there observing with infrared instrumentation provided by Caltech, and he also ventured south to the Las Campanas Observatory in Chile. His return to Cambridge in 1972 saw him writing up his observations before starting work as a Science Research Council Fellow at the Royal Greenwich Observatory, which was then at Herstmonceux Castle in Sussex. He stayed there for three years, collaborating in infrared astronomy with Dr Ian Glass (now of the South African Astronomical Observatory).

By coincidence, I, too, was at Herstmonceux for most of that period and, though I knew of David – 'the chap with the limp' – he managed to elude me completely. Partly it was because he was away a lot, but it also says something about the nature of RGO in those days that my rather pedestrian work on planetary orbits could keep me totally insulated from the exciting 'lunatic fringe' of infrared astronomy.

In 1975, David made his move to the Anglo-Australian Observatory, where he remained for the rest of his life. The AAO was then in its infancy, the 3.9-metre Anglo-Australian Telescope having been inaugurated only the previous October. The telescope itself is at Siding Spring Mountain, near Coonabarabran in rural New South Wales. David, however, was based with most of the scientific staff at the Observatory's laboratory in the leafy Sydney suburb of Epping. Again, he held a Science Research Council fellowship.

David took to Australia like a duck-billed platypus to a waterhole. He relished its enormous potential for walking, and became particularly attached to the spectacular Warrumbungles range where the AAT is located. The geology, flora, and fauna of the district held a special fascination for him. Eventually, that interest manifested itself in a little booklet that is not listed among his

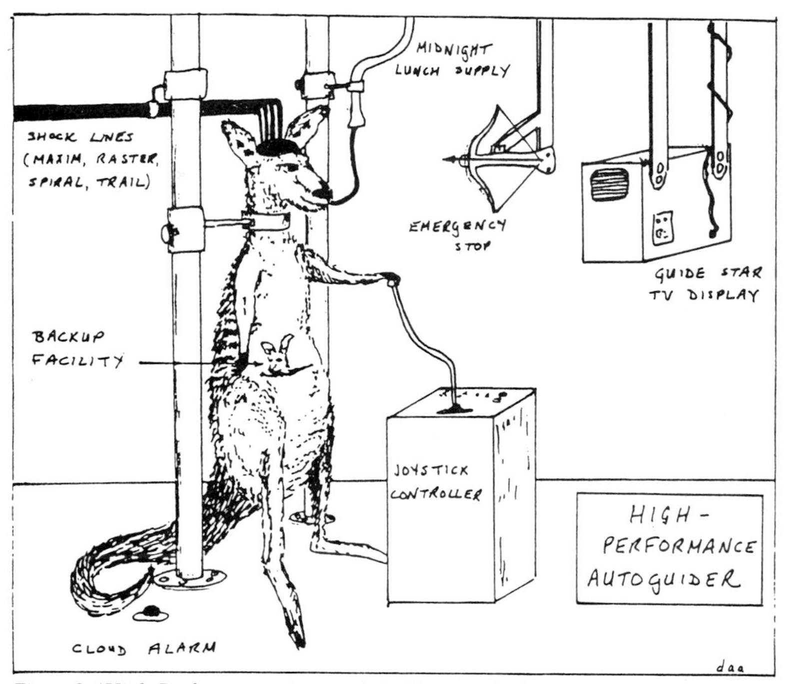

Figure 2. 'High-Performance Autoguider'. This David Allen cartoon graced the pages of the 1981 edition of the Anglo-Australian Telescope's Observer's Guide. *Though he had nothing but affection for Australia's most famous marsupial, his Goonish imagination could place it in some decidedly precarious situations.*

prestigious scientific publications, but which was very close to his heart. *Walks on and around Siding Spring Mountain* opened the eyes of a whole generation of visiting (and local) astronomers to the special appeal of the Warrumbungles.

It was in Australia, too, that the Allens' three children were born: Kachina, Andelys, and Alasdair. In 1983, David's position at the AAO underwent a unique transformation, in that he became the only member of the scientific staff to be awarded an indefinite, rather than fixed-term post. Though his new position freed him from the need to worry about future job applications, it brought new responsibilities too, and David found himself in the position of Acting Director of AAO more than once. He also had a term as President of the Astronomical Society of Australia. As might be expected, he carried out these duties with his customary accomplishment.

Instruments and inspiration

When David first joined the AAO, he was advised not to pursue his interest in infrared astronomy but, instead, to concentrate on observations in the visible region of the spectrum. The telescope

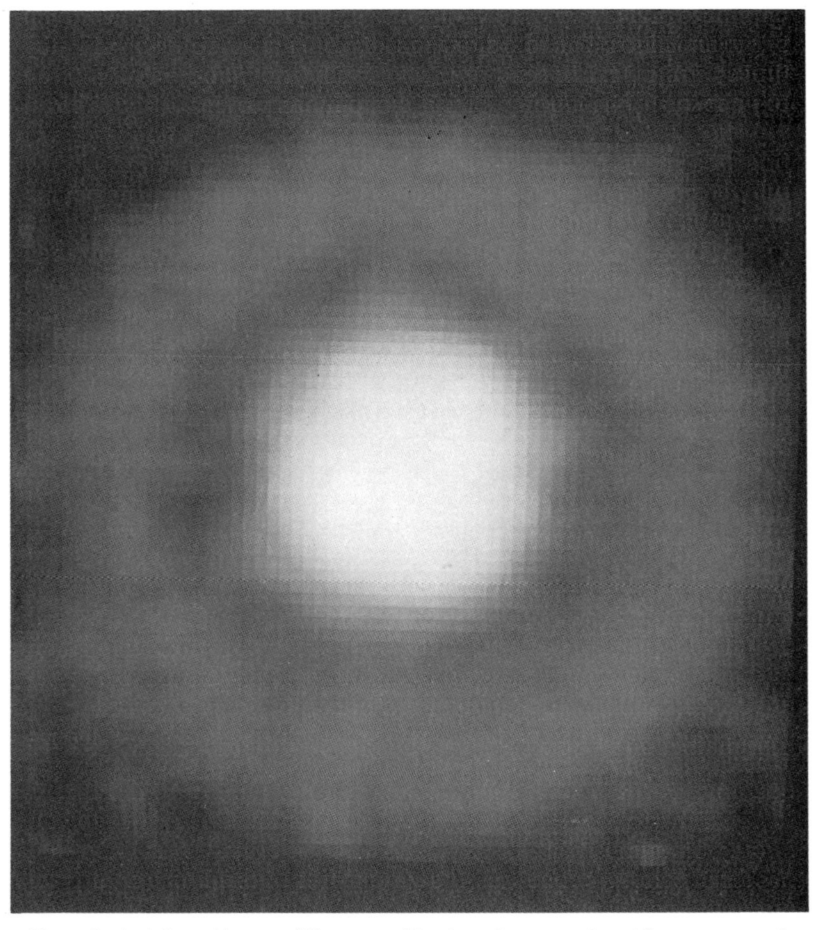

Figure 3. An infrared image of Uranus and its rings. It was produced from scans made with IRPS, the infrared photometer–spectrometer devised for the AAT by David Allen in the late 1970s. The image is a composite made at the two wavelengths of 1.2μm and 2.2μm. At the longer wavelength, the planet itself is dimmed by methane absorption, allowing the faint rings to be detected (Photo courtesy of David Malin, © Anglo-Australian Telescope Board, 1983).

site at Siding Spring was not considered suitable for infrared observations. Its altitude of only 1200 metres made it look like a mere termite-mound compared with the 4200 metres of Mauna Kea in Hawaii, then already recognized as one of the world's best infrared sites. And, in general, the atmospheric humidity was not as low as at the mountain sites of northern Chile. But David foresaw a future for the AAT in the near-infrared, at wavelengths shorter than 5μm. Here, the problems with water vapour, though still acute, are not as intractable as they would be at longer wavelengths.

Eventually, David persuaded the then-Director, Dr Don Morton, that a near-infrared instrument could be built and used on the AAT, and the result was IRPS – the infrared photometer–spectrometer. Commissioned in 1978, IRPS owed much to the genius of the AAO's engineering staff, particularly John Barton. It used a single indium antimonide detector to measure the infrared radiation collected by the telescope from individual targets in the sky. The detector was operated in such a way as to give it excellent sensitivity and stability. These attributes were vital, since they allowed IRPS to be scanned across areas of sky to build up infrared images – some of the earliest to be obtained anywhere.

David's role in the IRPS team was project scientist, providing astronomical guidance in its development and subsequent exploitation. This latter aspect gave him much satisfaction, for many astronomers observed with the instrument, and obtained new results from it. One such user was Professor John Storey of the University of New South Wales, who, in fact, collaborated with David in studies of the galactic centre. Professor Storey shared David's love of words. He had an aptitude for communicating with his colleagues in verse – to the extent that, on at least one occasion, he presented an entire paper at a research symposium in rhyming couplets! John recalls one particular IRPS software bug that drove him to write in the faults log:

> I set the gain to twenty;
> I thought that would be plenty.
> But every time I start a run,
> The gain re-sets itself to one.

David's reply, also in verse, is lost in the mists of time, but the bug was duly fixed! IRPS, incidentally, has recently found a new lease of life helping to monitor sky conditions at the South Pole for a possible astronomical observatory there; again, John Storey is playing a leading role in this work.

Following its success with IRPS, the AAO commissioned another infrared instrument called FIGS (Fabry-Perot infrared grating spectrometer). This was a collaborative venture between the AAO and the Australian National University. It was a true infrared spectrometer in that it used the prism-like properties of a diffraction grating to form an infrared spectrum on the detector, now no longer a single point, but an array of 16 indium antimonide cells. To permit it to work at wavelengths beyond 2μm (where heat radiation from sky, telescope, and instrument starts to become obtrusive), the entire spectrometer was cooled to the temperature of solid nitrogen (−210 degrees C). This advanced instrument started work in 1985, again with David's help.

IRPS and FIGS were eventually superseded at the AAT by the instrument that David will probably be best remembered for. IRIS, the infrared imaging spectrometer, uses a 128 × 128 array of mercury–cadmium–telluride detectors, and provides imaging at a range of scales, and spectroscopy in the 1 to 2.5μm wavelength region.

The story of IRIS blossoming from the seed of an idea to a finished instrument is an epic in itself, for the project hinged upon the release of an export licence for the US-manufactured detector array. These devices were originally developed for military thermal-imaging applications and still had strategically sensitive overtones. It was a great vindication of David's perseverance that the array was exported to Australia at all. I well remember feeling depressed when news of its release reached the AAO: it meant that work on *all* other instrumental projects – including my own FLAIR II – would be suspended so IRIS could be pushed ahead. It cost FLAIR II a year, but it was clearly the correct decision.

IRIS was commissioned at the end of 1990, heralding a new era of near-infrared imaging for the AAO, and keeping the Observatory in the forefront of infrared astronomy. Now, detailed images taken in different near-infrared wavebands could be combined to produce striking colour pictures comparable both in beauty and scientific content with the visible-light photographs made by David's colleague and close friend, Dr David Malin. Truly, infrared astronomy had come of age.

IRIS remains the AAO's flagship infrared instrument. It owes as much of its success to the skills of the AAO engineers who built it as

Figure 4. Explosion in the Orion nebula. This much-publicized image, taken with the IRIS infrared camera, tells of a violent outburst less than 1000 years ago from the star IRc2 near the centre of the nebula. Its discovery by David Allen was the subject of his last article for the Yearbook of Astronomy. *It appeared in the 1995 edition (Photo courtesy David Malin, © Anglo-Australian Telescope Board, 1993).*

to David himself, and he paid tribute to them when the instrument was awarded the Institute of Engineers' Bradfield Award in Sydney in 1993.

Celestial explorer

'Research remains my first love in astronomy, and I particularly relish the observational aspects of it.' David wrote those words near the end of his life as he looked back on a career literally crammed with scientific discoveries. Many of us, I suspect, would be satisfied to have emulated his achievements in astronomical instrumentation alone, but for David, they were merely a means to an end.

It was not just his own instruments that he used. His observations straddled almost the entire electromagnetic spectrum, from X-ray and ultraviolet satellite data, through the visible and infrared to long-wavelength observations with radio telescopes. Likewise, the objects he studied ranged from those on the Earth's doorstep – the Moon and Venus – to active galaxies and quasars in the farthest reaches of the Universe.

David's contribution in all these fields of work was significant. A bibliography of his publications stretches to well over twenty pages. Arranged chronologically, it emphasizes the breadth of his interests, the titles darting from one topic of research to another. Infrared photometry of Saturn; photometry of variable stars; Wolf-Rayet stars; asteroids; planetary nebulæ; peculiar emission-line stars; Seyfert galaxies; carbon stars; X-ray novæ; water vapour in molecular clouds; dwarf galaxies; symbiotic stars; dust shells, quasars; the rings of Uranus; the galactic centre; supernovæ; comets; superclusters of galaxies . . . the list goes on and on, in a breath-taking rollercoaster ride through modern astrophysics.

Let me try to sketch just a couple of highlights. To find others, you can do no better than to look back through previous issues of the *Yearbook of Astronomy*. There, you will usually find an account of some aspect of David's work, written in his own inimitable style.

Lifting the veil on Venus

The planet Venus is depicted in the tea-packet picture cards without no information other than details of its orbit. That is because, in 1955, when the cards were issued, little else was known about Earth's sister planet. Thick clouds perpetually hide its surface from the gaze of visible-light telescopes. Perhaps that mystique was what sparked off David's love-affair with Venus; in any event, he began in 1983 to make serious infrared observations of the planet.

Using IRPS to scan the dark side, he discovered that patterns of clouds in the planet's lower atmosphere can be detected, silhouetted against a 2μm background glow originating close to the surface. It was the first time that Venus' turbid atmosphere had been penetrated by ground-based observation, and it allowed detailed study of the planet's cloud circulation.

Eight years later, using the newly-commissioned IRIS, David extended his work to other infrared wavelength-windows in the

Venusian atmosphere. He and his collaborators found airglow emission analogous to Earth's auroræ (though of different origin), and realized that it could form a valuable probe of circulation in the tenuous upper atmosphere. Even more spectacularly, they discovered that at certain wavelengths, the Venusian surface itself is visible by its thermal emission. The data recorded from these observations complement those from other, more exotic (and expensive!) sources like the Magellan radar-mapping spacecraft. David counted this work among his most important, and was still probing the planet's secrets when he died.

Starburst at the galactic centre

Like the surface of Venus, the centre of our Milky Way galaxy is hidden from us at visible wavelengths by opaque clouds. This time, the absorbing medium is interstellar dust, better thought of, perhaps, as smoke. Herein lay another of David's fascinations, for this dust can be penetrated by infrared radiation. The galactic centre was thus a natural target for his curiosity almost from the beginning of his career. The fact that it passes directly overhead at the AAT during the long nights of winter was a bonus.

Working with Dr Michael Burton (now of the University of New South Wales), David used IRIS to obtain images of the galactic-centre region in various different wavelength-bands simultaneously. Maps were produced showing the distribution of ionized (*i.e.* excited) and molecular (*i.e.* cool) hydrogen, together with helium.

Allen and Burton interpreted the results as being due to a group of hot stars around the galactic centre, created in a burst of star-formation a few million years ago. This flew in the face of the conventional model, which envisaged no new star-formation as recently as that. Aficionados of that picture sprang to its defence, proposing new mechanisms – like collisions between the remnants of very old stars – that would account for the apparent population of young stars. But later work by the two AAO astronomers (again using IRIS) brought new support for their theory.

The new observations revealed that a wide range of star masses is present in the group of hot stars, which is exactly what would be expected from the 'star-burst' model. It is very difficult to explain by collisional or other theories. Thus, important new evidence has been gathered on the galactic centre's eventful history, and the astrophysical phenomena that have shaped it.

The people's astronomer

I was originally going to call this article 'The Astronomer's Astronomer'. It is a description that fits David well, for his interactions with other astronomers in all fields of the subject were usually very successful. True, his manner sometimes appeared brusque – arrogant, even – and he certainly didn't suffer fools gladly, but he always went out of his way to help where he could. He had a particular regard for students and young scientists, never forgetting the difficulties that have to be faced in making a career of scientific research. His generous encouragement will be remembered by many.

Perhaps Dr Russell Cannon, the AAO's Director, put his finger on David's appeal to other scientists when he described him as 'always original in his thinking' – a view that echoes the opinion of David's Ph.D. supervisor almost thirty years ago.

However, to paint David only among his astronomical colleagues would be to leave the portrait half-finished. Next to his family, the great love in his life was communicating his science to the world at large. Public lectures, radio and TV interviews, articles, essays, and books; all were part of the daily routine in a schedule that would leave most of us wondering when we might manage to get some sleep. And it was not only astronomy that David wrote and spoke about. He had a lifelong interest in the natural world, a fascination with every facet of creation that showed up in most of his activities. He once found himself President of the Fossil Club of New South Wales, for example.

David had a standing invitation to appear on *The Sky at Night*, and that unruly forelock of his – the one that always seemed to need pushing back out of his eyes – became quite familiar to regular watchers of the BBC's late-night TV astronomy programme. But it was his writing that drew the widest admiration. He had a real flair for bringing science to life, writing in a style that Robyn Williams of the ABC's *Science Show* described as 'almost poetry.'

I remember, years ago, watching David at work at a keyboard in the AAT's control room. He was unaware of my presence, and the article he was writing simply flowed through his fingers in a continuous stream, demonstrating a clarity of thought that was, to me, little short of stunning. 'How does he *do* that?', I wondered – and still do, as I agonize over every word and phrase I write, groping for the means to express myself in some intelligible way . . .

In his profession, in his writing, in his speaking, David truly was the Universal Astronomer. He is much missed by all of us who knew him. I firmly believe that, had his life not been cut short, he would eventually have become in Australia what his first mentor – the Editor of this *Yearbook* – is in Britain: astronomy's national treasure.

Postscript

David is one of a number of astronomers of his generation to have succumbed to cancer in recent years. It saddens me greatly to have to tell you that another friend in Australian astronomy is now fighting bravely against a brain tumour of similar voracity to his.

This person is not someone of David's generation, nor my mother's, but a vivacious young woman of considerable personality and talent. Her research has taken her to the brink of infinity; now she measures her own life-expectancy in months. I saw her during a brief visit to Sydney a few days ago. She looked radiant, but the five purposeful-looking tablets she took with lunch betrayed the seriousness of her illness. It was as much as I could manage to keep the tears at bay. All who know her hope for a miracle.

What are we to do about this awful disease? Should the great thinkers in the fundamental sciences like astrophysics and cosmology – the David Allens of this world – forsake their studies and add their intellectual fire-power to medicine's onslaught? Should they re-direct their talents towards eliminating the gliomas, sarcomas, and melanomas that indiscriminately rob us of our friends and loved ones?

In spite of everything, I have to say that I believe the answer is no – and I think David would have agreed. It wouldn't work. The efforts of those special people whose thoughts carry them beyond the limitations of human frailty are vital to our understanding of the Universe. And science must explore every aspect of Nature if we are to reveal the complete picture.

In the end, our most gifted scientists must be allowed to follow their calling. For who knows just what we might learn from the Cosmos that could one day help to abolish suffering here on Earth?

The Planets of the Virgo Millisecond Pulsar

PAUL MURDIN

There are enormous numbers of stars. There are several thousand visible to the unaided eye and 10 thousand million stars in our Galaxy, most of them massed in star clouds in the Milky Way. For centuries people have wondered, since there are such vast numbers of stars, whether there are correspondingly vast numbers of worlds in orbit around them, in the same way that our own planet, Earth, orbits our star, the Sun.

Most people view planets as the results of natural processes. They argue that our world and its eight fellows in the Solar System are just the few planets that we know and that there are many others. Because planets are small, and have no light of their own, they are difficult to see. So, until now, no planet has been detected outside of our own Solar System.

Now it seems practically certain, from recent measurements by a radio astronomer, that three planets have now been detected in orbit around a pulsar.

Pulsars are spinning neutron stars. Although the same mass as the Sun, they are very small, say 10 km radius, the size of the orbital motorway, the M25, around London. Pulsars spin at high speeds, typically once to 1000 times every second. Radio astronomers see pulsars because they flash radio waves. Each pulsar has a radio beam fixed on the rotating neutron star, pointing out like the light beam of a lighthouse. As the neutron star rotates, like the lighthouse mechanism, the beam flashes when it sweeps across us. A pulsar is like a lighthouse but rotates much faster, typically several times every second. The rotation makes the radio waves pulse – that is why pulsars have the name they do.

Pulsars are made in supernova explosions. In a supernova, the core of a heavy star collapses and releases a lot of energy, which astronomers detect as an explosion. The collapsed core shrinks from several thousands of kilometres in size to just 10 km radius; the collapse of the core makes a neutron star. The neutron star picks up spin and so it rotates quickly, at birth maybe 100 times per second. The collapse compresses the magnetic field which is trapped in the core, so the neutron star is highly magnetized and that is how the radio beam is generated. The best-known pulsar is the one in the middle of the Crab Nebula; it rotates 30 times per second, flashing not only radio waves towards us but also light waves. The reason why the Crab has a pulsating neutron star in the middle is that it is an explosion produced by a supernova in the year 1054.

Pulsars energetically radiate energy away and slow down. In fact, their spin energy feeds the pulsations. Over a period of some 10 million years, a typical pulsar changes its spin rate from 100 times per second to once per second. Radio astronomers see hundreds of pulsars slowing down.

There are a few exceptionally fast pulsars; it is believed that they are formed in binary star systems, from a slow pulsar which has been speeded up. If there is a supernova explosion in one of the two stars in a binary star, it might be that the supernova forms a pulsar which remains in orbit around the other star. The pulsar slows down as usual. But then the ordinary star lives out its life and grows in size, so its atmosphere dribbles in a stream on to the pulsar. The impact of the stream on the surface of the pulsar makes it spin faster again.

The pulsar speeds up maybe to 1000 times per second – its period is measured in milliseconds. Because it is rotating quickly, the pulsar has become very energetic again, and it heats up and evaporates the ordinary star; or maybe the ordinary star explodes as a second supernova, perhaps disrupting the binary system, and producing a lone millisecond pulsar. Radio astronomers see several millisecond pulsars, and believe that they have been formed in this way.

It is a bit surprising that such fast pulsars are actually old – usually pulsars start fast and go slower, so we think of the slow ones as old, say 10 million years old. But the really fast ones are even older, if they have been through this speeding up process, say a billion years old.

These fast pulsars are very attractive to radio astronomers because they can measure their periods very accurately indeed, revealing small irregularities. Fast rotating neutron stars are very accurate clocks, but they can appear not to tick exactly on schedule because of the very subtle effects. For instance if a pulsar is in orbit around something, its pulses arrive late when it is on the far side of the orbit and early if it is on the near side.

There is quite an industry of radio astronomers measuring pulsar periods accurately, to study the subtle effects of the slight irregularities. One of these astronomers is Alexander Wolszczan of Pennsylvania State University; it was he who discovered planets in orbit around a pulsar.

In 1990 Wolszczan was working at the Arecibo Radio Observatory in Puerto Rico, when he was lucky enough to have the exclusive use of the telescope for a few weeks when it was closed to outside users for structural alterations – although it was still able to survey the sky overhead. He discovered a millisecond pulsar catalogued 1257+12 in Virgo (in the strip of sky overhead Puerto Rico). The pulsar is at least 1500 light-years away, perhaps twice that. It has a period of 6.2 msec – it beats 161 times per second. If you could hear it, it would sound like a lowish note on the piano (G below mid-G).

Wolszczan monitored the pulsar for a couple of years and found it was not quite regular. Sometimes the pulses came almost a whole beat too early, sometimes almost a whole beat too late. In musical terms the note was a vibrato. There were two vibrato periods, of two months and three months respectively.

The effect was very subtle but early in 1992 Wolszczan had watched for long enough to get the behaviour clear, and he announced his interpretation. The pulsar was in orbit around something else – not just one thing in fact, but two. The pulsar was in orbit around the common centre of mass of the system. It was this small displacement, about 1.5 light milliseconds (5000 km), that caused the pulsar's pulses to arrive, first early in their schedule and then late. The two additional bodies in the system corresponded to the double vibrato of the pulsar's beat.

The periods of the two additional bodies as they orbited the pulsar were two months and three months respectively. The great surprise was this: the other two things were not as massive as the pulsar – this was not a triple star system. The other two things had masses which were more like the masses of planets (three or four times the mass of the Earth) and they were separated from the pulsar by distances which are reminiscent of the Earth–Sun distance. The triple system was like a small solar system, a sun and two planets with near circular co-planar orbits.

There was more to come: at the International Astronomical Union meeting in the Hague in August 1994, Alexander Wolszczan announced that he had found a third planet in the system from a third subtle effect which had taken some time to emerge.

The new planet orbits closest to the pulsar and is more the mass of the Moon than the mass of the planets. There are still some unexplained changes in the pulsar's ticking which could be caused by one or more planets with long periods orbiting the pulsar at distances like Mars and Jupiter's distances from the Sun. Time will tell if this is the case.

The planets of the Virgo millisecond pulsar

		Period	*Mass*	*Separation*
Pulsar		6.2 ms	1.4 × mass of the Sun	
Third object	A	25.3 days	0.014 × mass of Earth	0.19 AU
First object	B	66.6 days	2.8 × mass of Earth	0.36 AU
Second object	C	98.2 days	3.4 × mass of Earth	0.46 AU

There was a false alarm about a planet round a pulsar in 1990, and astronomers have been a bit wary about subsequent claims of the same sort of thing. But these observations show that a mistake is not likely. When the first two planets were discovered, theoreticians predicted an effect in the timing measurements which would only be seen if there were indeed real planets in the system. They realized that the orbits of the two heavy planets interact. Every three revolutions of planet B correspond to two of C. So if at a certain moment the planets are close to each other, as the inner one overtakes the outer one, they pass each other closely again 200 days later. As B and C pass closely, they give each other a gravitational tug. Because of the resonance effect, the tug builds up, altering the positions of the planets and therefore of the motion of the pulsar around the centre of mass of the system. Resonance effects like this are seen in multiple satellite systems around some planets in our Solar System. The effect of the tug on the pulsar's planetary system has now been seen in the data. It is very hard to think how something different from planetary dynamics could mimic this effect.

It is interesting to speculate what these planets look like. They are lumps of cosmic stuff of the same size as planets and they have had a billion years to grow as normal planets do. I think they could well look like normal planets. They could well have a surface as the

Earth does and the Moon does. The planet A is the lightest and the closest to the energetic pulsar so it won't have an atmosphere, for the same reason that Mercury doesn't have an atmosphere, being so close to the Sun. Planet A probably has mountains and maria like the Moon. It probably does not show many craters, because cratering in the Solar System seems mostly due to a unique event when a planet broke up, and bits fell on to other planets and satellites (and there is no reason to expect this to be usual in all planetary systems). My best guess is that A looks like the Moon without craters.

Planets B and C are heavier than the Earth and Venus, but they are only a quarter of the age of these two Solar System planets. The surfaces of B and C could well look like the surfaces of Earth and Venus in their early days, with copious volcanic activity. The volcanoes will emit gases, which will be held back by the relatively strong surface gravity of B and C, and they could have a denser atmosphere than either Earth or Venus. However, maybe the magnetic wave energy of the pulsar, so nearby, will quickly sweep any atmosphere away. Planets B and C will probably have magnetic fields like Earth and Venus, and the interaction of these fields with the pulsar's magnetic field will be interesting – maybe there will be brilliant auroral-type phenomena, not atmospheric auroræ if B and C are bare, but surface auroræ with the rocks fluorescing and showing fantastic discharge phenomena!

The origin of the planets around the Virgo millisecond pulsar is unknown. One thing seems fairly sure: it does not seem possible that planets formed at the outset of the pulsar's life could survive being in one or two supernova explosions and an interacting close binary star. These planets are probably not so-called first generation planets.

Maybe they are a chance phenomenon, generated by a coincidence. Perhaps the Virgo millisecond pulsar was once a lone pulsar and passed near a white dwarf, through its planetary system – an old solar system of the sort our own is destined to be. As the pulsar passed through, it may have stolen three planets. This seems a pretty unlikely scenario, because it needs the pulsar to pass close to the white dwarf. What is more, the stolen planets will be captured into very eccentric orbits, and the orbits of the three planets orbiting around the Virgo pulsar are circular.

There are more natural explanations which are generic to the history of the millisecond pulsar. Maybe they are second generation planets formed directly around the pulsar. One thought is that after the millisecond pulsar was formed in its binary star system, it evaporated most of the companion. The scraps that were left formed first a planetary disk and then planets. Another possibility is that the companion exploded as a supernova and some bits of the explosion fell back and got caught by the pulsar, again making a disk and planets.

It could well be that radio astronomers will soon discover more planets orbiting other millisecond pulsars – it just takes time to see the effects of the orbits! The discovery of the first planetary system to be found outside our own, and the prospect of finding more, reinforces our ideas about the formation of planets. They are indeed natural phenomena, produced from a disk of cosmic stuff in orbit round a star, after a billion years. Just to find one more planetary system in such different circumstances from our own Solar System must mean they are indeed common throughout the Universe.

Annular Eclipses

MICHAEL MAUNDER

Why Annular eclipses? It is a bit simplistic to reply, 'because they are there', as the reason why you should observe them. Annular eclipses are as rare as totals, and the science to be got from them can be quite as important on occasions. In the next few years this is very true, because some very interesting phenomena occur which can best be seen during an annular event. Some of these interesting events will be reviewed here.

Even if the reader considers annular events as a little 'boring' or at best 'second best', there is a very good reason why you should think again. Such events are the best training or practice you are likely to get for a total! If it all goes wrong you can at least brush it all under the carpet and pretend you were not really all that bothered. However, there is a serious side to this introduction, which could appear flippant at first.

It is this question of practice. All the techniques needed to follow an annular eclipse are the same as used in ordinary solar observations. If you are a regular solar observer, then why not use that expertise to chase an annular eclipse? When a total one comes along you have something substantial to fall back on, particularly experience of the excitement of the moment. But there is an even more serious point to be considered.

If you are even just a little bit tempted to chase an annular eclipse, why not turn that interest into becoming a serious solar observer?

As I have said already, the equipment and technology can be the same for both astronomical pursuits. What has to be said, can equally be applicable for a much longer interest in solar studies, and that has to be good news for astronomy in general. In these days of night-time light pollution, the Sun is readily seen every clear day, for the whole time of sunrise. If the techniques are modified a little – no filters, for instance – a lifetime interest in lunar work can be pursued during the dark hours.

So, annular eclipse studies are really only a specialized spin-off of either or both regular lunar and solar observing. The real spin-off is in the travel.

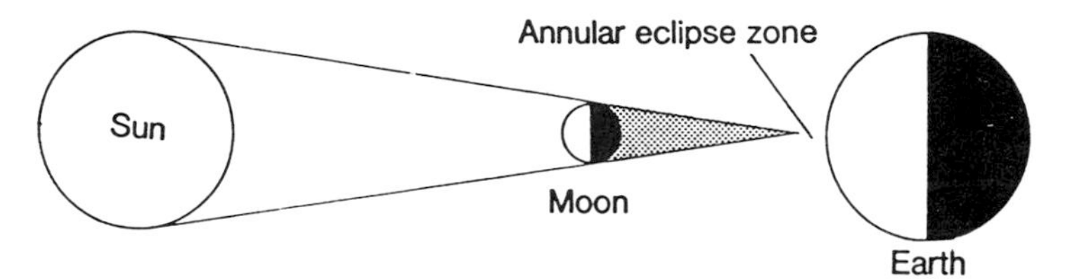

Nobody who has seen any form of eclipse, whether it be a total, or a lunar event, even an occultation, forgets it. Many become addicted to these rare events. The addiction takes many forms and can lead to a great deal of satisfaction. You just have to take the trouble to be fully prepared.

Because all forms of eclipse-chasing require a similar philosophy, I am taking this opportunity to use a similar layout to the piece on total eclipses in the *1990 Yearbook* in bringing that discussion up to date. We start with the crucial question of safety.

Safety Warning on Filters

Annular eclipses, like regular solar observing, are not like total solar eclipses. *You cannot gaze at the important bit with an unprotected eye. You have to be on your guard all the time.*

This warning cannot be repeated often enough!

No filter can be entirely safe when placed behind the main mirror or lens, as the only filter before the eye. YOU HAVE BEEN WARNED! Never, ever, consider them, whatever the source, or type, and still, never ever, regardless of the reliability of the person recommending them.

Things can and do go wrong in real life. The politest description of the Universal Law which goes 'If anything can go wrong, it will, and in the most inconvenient way', is Spode's Law. You only have one set of eyes to play with, and whatever your attitude to animals might be, I am sure that you do not wish to swap a free-range/willed dog to being your slave in getting you to the next (for you) words-only astronomy meeting. Guide dogs are too valuable to be wasted on someone not blind through no fault of their own, I cannot make my point much stronger.

Filters placed in front of the main optics – whether eyeball, telescope or camera – must have the right properties. Virtually all the eye damage seen at hospitals arises from radiation outside the visible spectrum.

Ultraviolet radiation (UV) is well known for its damaging effects such as sunburn. For eyes, the main risk in solar observing is much the same. The cooking in the eyeball can lead to similar effects to sunburn (you are unlikely to get enough radiation to end up with cancer), but cataracts on the cornea are a strong possibility. They will not take decades of sunbathing to manifest; periods as short as months have been noted.

UV radiation will pass through a surprisingly large number of visually opaque filters. The stuff is also blocked completely by some (visually) transparent materials, which means that you must use only solar filters sold for that purpose.

Infrared radiation (IR) is even more prone to pass through visually opaque filters, and almost every photographic filter of the 'Wratten' type, including neutral density ones. IR is pure heat radiation. It concentrates on the retina or in the eyeball fluid, literally cooking it. Irreversible vision loss occurs in the areas affected, and shows up immediately.

Solar Filters

MYLAR

Mylar filters have become the most popular type for solar work in recent years. Mylar is an extremely tough plastic film. The type to use is 10 microns thick, coated with a very thin layer of aluminium metal. For visual work, the stuff coated on both sides is by far the safest, whereas single-sided is better for photography because of the shorter exposure times this allows.

The aluminium coating absorbs both UV and IR radiation and transmits only a small amount of visible light, mainly blue. Properly prepared and selected, Mylar filters are safe enough for everyday use.

Mylar film must never be confused with 'Silver Paper', which is a thin sheet of aluminium metal, nor with glitters. Glitters are similar plastic sheets but with a silver or metallic ink printed on to them. They are also becoming very common in the packing industry, and are very dangerous indeed because they pass too much damaging radiation.

Only use Mylar from a reputable source!

Mylar itself is extremely tough, but the aluminium coating is not. Each piece must be inspected for pinholes. This does not mean just before leaving on an expedition, but each and every time before you put your eyes to the optical endpiece. Pinholes transmit a lot of damaging radiation – that is, after all, how a pinhole camera works – and also cause image degradation due to halation. This is seen as lowered contrast and sharpness.

Inspect the surface regularly for unevenness. Scuffing and other abrasions eventually lead to local highspots and pinholes. The dangers are obvious and image degradation can be appreciable. Some slight rucking or unevenness will not cause image degradation. On the contrary, some authorities positively recommend not stretching the film too tightly.

If a pinhole or hotspot is seen, replace the filter immediately; the stuff is cheap enough.

The main problem with Mylar arises from a surprising reason which is rarely mentioned or discussed. The film is 'poled', which means it has a marked polarization. You can see this by holding up a piece to sunlight. The Sun's image will often have 'wings'. These rotate as the piece is turned round. The effects really become serious with some types of camera viewfinder, particularly auto-focus where circular polarization is needed. Before attempting to photograph anything important, check focus by rotating the filter and always double check with test exposures.

Mylar film can be obtained in a very wide range of densities, and this leads to a few problems of its own. There is a tendency to use the lighter versions for photography and forget the visual dangers during focusing and regular visual monitoring. For the safest working, use the denser products always. However, there is considerable merit in keeping exposure times short, particularly with the longer focus lengths needed in annular eclipses and for granularity studies, etcetera in routine solar surveys.

One simple dodge is to use two thicknesses for visual work and for focusing. Although the image will be degraded a little, a point of best focus is quickly found. The other is to use a stop, although this can shift the focus quite a lot with 'cheaper' optics.

ALUMINIZED GLASS

These filters are increasingly popular, and give every satisfaction apart from cost.

They are little more, in essence, than a rigid version of Mylar, and are made in the same way by vacuum deposition of aluminium. Any company offering an aluminizing service for a normal astronomical mirror should be capable of turning out one, but many do not have the necessary quality control to ensure the absence of pinholes and hotspots. Only buy from a company offering some track record of reliability.

The same points made about Mylar apply without the problem with 'poling'. The important difference is that the optical properties can be that much more rigidly specified in manufacture and sale, and you pay accordingly.

Aluminized coatings can be laid down on virtually any size of glass, so the limiting cost often comes down to the price of the glass itself. This must be of the highest optical property, and sizes above 100-mm diameter are prohibitively expensive. Fortunately, we do not need pieces bigger than this for most occasions since we are (mainly) concerned with portable set-ups. A popular size is 72-mm diameter, which covers many of the cheaper-end-of-the-market mirrors lenses, ideal for cabin baggage.

Treat these filters with respect and they should last a lifetime. Avoid fingers and any contact with salt (seaside) or other corrosive atmospheres. Mount them with the aluminized surface innermost, so that the glass itself acts as a nice protection.

Like Mylar, the image tends to be a strong blue although some makers claim a neutral tinge. If you can afford it, go for some surface coating on the aluminium as an extra protection over the years.

Consult the trade manuals and magazines for current prices and availability.

INCONEL

This metal coating is a special variety of stainless steel on optical glass. At present, these special solar filters are only readily available in USA, most UK companies declining to produce. However, they can be obtained from almost any reputable optical supplier under a different name as beam splitters and neutral density filters. They are the best available and are produced to extremely precise optical specifications. The drawback is the cost.

The reason is that beam splitters and neutral density filters are intended for very precise optical work. The other snag is that virtually nobody makes these above 50-mm square, 25-mm being the norm. If this is big enough, they have to be the ones to use because of their pedigree.

Other varieties of stainless steel filters are produced by at least one American company. These are made as large as you can afford in a variety of densities for both visual and photographic work.

Every solar observer should aspire to own at least one of these, if only for the peace of mind of specified optical properties and durability.

SILVER

Including this type of solar filter might appear to be out of place because it is generally assumed they are not available. Not strictly true as at least one company, Balzers, offers a pure silver filter for beam splitting and as a neutral density filter, just what we need. These are specially coated for durability. The snag is the same as Inconel in existing only in small sizes, and cost.

The main advantage of a pure silver mirror as a filter is the perfectly neutral image colour. Another is that it is even better than Mylar in filtering out ALL infrared radiation, and thereby safer.

It was not so long ago that all astronomical mirrors were silver. The cycle is repeating itself, and the latest generation of telescopes intends to have that metal coating because of its superior reflective power. Modern surface coatings ensure a respectable lifetime. Whilst I do not suggest that professional observatories sell their expertise for our purposes, I guess it will not be too long before pure silver mirrors are sold in the amateur market. Meanwhile, consult some of the standard textbooks on silvering and see if it is worth while dabbling. I think so for the more dedicated solar observers, but the recipes are outside the scope here.

'BLACK' BLACK AND WHITE FILM

This is the 'poor-man's' answer to silvering glass. It is safe to use if dark enough and gives much better colours than Mylar. So why is it not seen more often? The reason is that the image is rarely easy to focus and has low contrast due to halation arising from the Callier Effect. This is light scattering by the (appreciable) size of the silver particles in the emulsion. Those in the know get over the Callier effect to a large extent by using lith or line films which have very fine-grained emulsions of highest contrast. They are also available in sizes bigger than any normal telescope, and are made with a backing of that magic stuff, Mylar. Provided these films are developed and fixed properly and checked over for pinholes, they can give images almost as good as Mylar film and Inconel.

The great advantage is that it makes an excellent back-up if all else fails, and you can have a whole range of filters prepared in advance with a range of densities to cover every conceivable cloud cover. If you regard monochrome film filters in this way they are definitely not 'second best', and every solar eclipse traveller should acquire some, if only to use as a simple naked-eye visual filter.

'BLACK' COLOUR FILM

Any form of colour film, no matter how dense it looks:
NOT TO BE USED UNDER ANY CIRCUMSTANCES

WELDING GLASS

Somewhat undervalued until recently, this material can be got in a very wide range of densities, called 'shades'. The direct viewing shades, in the UK, are 13 and 14, and these are also right for most photography, although 12 is often better.

Perhaps the reason for its lack of popularity is in finding a retail outlet. Another is the difficulty of getting perfect optical flatness in larger sheets, and the sheer weight of them.

If you do have a source, it is best to accept them as they are, which is rectangular, and not to have them trimmed circular as this often generates all manner of stresses and strains which show up in the solar image (not so serious for direct visual work). Depending on the supplier, the image colour can be a nice neutral to a bilious green. Fortunately, the colour bias can be eliminated in photographic/video reproduction.

Some workers with very precisely collimated optics, mainly mirror optics, might find welding glass more of a problem than with glass lenses. The reason is the risk of secondary images from the widely separated welding glass surfaces. Try it out in both optical systems before rejecting completely.

Welding glass is often a good choice for the less critical optical applications, but where safety is paramount. Consider these filters for binoculars used in routine scanning and monitoring before more precise work is done through the main optics. Welding glass can be Araldited into a rigid mount, which can be securely fixed to the optics. Welding glass is very difficult to break because of its main application under hazardous conditions, also its thickness, and it still works well enough for our less critical purposes even when scratched! It has to be the filter of choice for safety as it comes with internationally agreed standards from all sources.

WRATTEN NO. 96 FILTERS

Properly kept, these can be optically as good as any. They are probably the best choice for pre-set or rangefinder cameras, that is, where the image is not seen directly. You just run a series of test exposures, using the wide range of densities offered, and the camera speeds, and you are calibrated for all time.

Remember that these neutral density filters are totally transparent to infrared radiation. Images produced through a Wratten 96 filter by direct means, such as a normal SLR viewfinder, MUST NOT BE VIEWED WITH THE EYE UNDER ANY CIRCUMSTANCES. So much infrared comes through, that the filters might just as well not be there, because of the medical damage they can do.

The reason why such filters let through infrared is very simple in historic terms. They were used for light PROJECTION purposes. So, if the infrared was absorbed, the filter would have burst into flame!

This is bad enough for the Mark I eyeball, and irises and other delicate bits in a camera shutter. You must think in terms of shielding the filter and the optics behind it whenever possible to keep this heating effect to a minimum. Where you must be very careful indeed is in the modern technologies of video/CCD. The detectors are extremely sensitive in the infrared. You might think them safe enough behind a Wratten 96 as no visual image is seen, but the full blast of solar infrared heating is still going through. It is best to forget Wrattens for video/CCD work in case you blow the electronics.

SILVER PRINTED PLASTICS

No matter how suitable it might appear to be:

NOT TO BE USED UNDER ANY CIRCUMSTANCES.

Which Eclipse?

Keen eclipse chasers will have a list for many years ahead. Some of the more interesting annulars are reviewed in more detail later.

For the rest of us start planning years in advance. Nature waits for nobody. Unless you are at the right place at the right time, forget it.

Finding the money is really a question of priorities as few are wealthy enough to chase each and every eclipse. Study where the next annular eclipse is taking place, and when. Then decide which eclipse best fits into your other plans. Having got the really difficult decision to go out of the way, now is the time to double-check that you are chasing the right eclipse.

Many eclipses take place in tropical countries. Anyone with a fear of 'creepy crawlies' will have to make their personal choice between the attraction and that disincentive. If the phobias are too high, a temperate climate could still be a bad choice, in the midge season. It will also inconvenience anyone with a severe allergy to insect bites. Similarly, freezing conditions or high altitudes are risky for certain medical afflictions, as was discovered in Chile in 1994.

It is worth repeating the advice I gave in 1990 on some of the basic issues:

> A few hours spent in a library checking travel agents' brochures is always time well spent. It will confirm or deny the climatic conditions, AT THE TIME OF YEAR YOU INTEND GOING. Is it the rainy season, for instance?
>
> Moving along the eclipse track to another place or country can make all the difference. If 'where?' is not too important a consideration, it boils down to deciding to have a good holiday in a country you would like to go to sometime.
>
> Some decisions on the country are made on very basic issues. Age and infirmity are something beyond control. If you have a problem, the hospital facilities must be adequate in case of emergency. Conversely, fit people may have important vaccine allergies. Yellow fever vaccine is based on eggs, and many cannot take it. Local regulations might require a certificate, so that country is out unless an alternative is found. Politics and ethics are also beyond control. The old saying 'When in Rome . . .' cannot be overstated. Anyone with strong objections to a régime or conditions in a country is well advised not to go. Make your protest some other time and place. I for one do not want a rifle butt across my head because you forget that basic axiom and started a riot.
>
> Once a short list of countries is drawn up, the viewing conditions should be the deciding factor. Everyone wants the best skies, but the intended type of work might swing the choice. Photographers have to consider the backdrop – paddy fields or an icy hillside.
>
> Without a doubt, the best way to get to an eclipse for the first time is to use a specialist travel agent. There are many of these advertising. Only consider alternative methods if you know the country, are an experienced traveller, or value the travel experience more. In many countries off the beaten track, the choice of hotels and accommodation is often limited, and snapped up by professionals very early on. The specialist firms also have some influence when it comes to selecting and varying the viewing site. This can be very important. A friendly police force can work wonders securing necessary travel permits or supplies.

This factor turned out to be very important for the total eclipse in Chile in 1994. Had it not been for the military taking us under their wing, the altitude sickness cases, site selection, and travel ease would have been all that more difficult. With a numbers restriction, private parties could have found the conditions almost impossible to deal with the unpredictable weather situation prevailing at the time. It is interesting that what I said in 1990 turned out to be the case in 1991 and 1994:

> Seasoned eclipse chasers often prefer to make their own arrangements. However, it is interesting that many such find their way to near the 'official' groups in the end! In the case of eclipse tracks entirely over the sea, there is little option but to take what ship is on offer. The choice has to be made on cost and who is likely to be on board. Ship travel is often the answer to those with 'problems' with certain countries. Getting to the country can often be only the beginning of the story because of the unpredictable weather. The intrepid eclipse chaser must be prepared to uproot all their gear and prejudices and simply take off to the hills as nature decides. The folklore of eclipse chasing is littered by true stories of city dwellers totally unprepared mentally to make the effort to put up with a non-air-conditioned hut for a single night.

Forget foreign travel to exotic places if you have never left home comforts before, and can't bear to be without them.

> If you are not prepared to rough it at the last minute, arrangements are always made by the specialist travel agents to have at least a reasonable outpost of civilization in the track to cater for that.
>
> Such a pity to spend so much time, money, and effort getting so far without that final mental will to get the best viewing site.
>
> Real planning can now start in earnest.

Photography or Viewing?

Do not be too ambitious, certainly not on your first trip. Even experienced photographers, myself included, make mistakes. The best advice to offer newcomers is to regard photography whether with a 'steam' camera or the latest video technology as a bonus. Go prepared to view the eclipse with nothing more than the naked eye, filtered, of course.

The great advantage of naked-eye viewing is that you have no clutter and last minute changes of plan are easy. You also have time to stand and stare, and even though I take a lot of equipment with me, the most important part is just standing and staring. This is particularly important in studying the landscape around you. In your mind will be the everlasting pictures, and no photograph ever replaces that for vividness.

Photographers should always come mentally prepared to ditch all their cherished plans (equipment sometimes!) and fall back to simple naked-eye viewing. Someone, somewhere is bound to get a good picture which you can keep as a souvenir. Snapshots of the site and people there can also have some value later, particularly if anything unusual happens.

Safe Viewing

By far the safest viewing method is projecting the Sun's image on to a screen. It is also the safest photographic option. The *Sun's image must never be looked at through any form of camera viewfinder without proper filters*. Some extremely fascinating views and pictures can be made by applying the 'pinhole camera' effect. Just position yourself under a tree so that you can see the dappled shadows thrown by the Sun shining through the gaps between the leaves. These gaps act just like pinholes, and as the eclipse progresses, the shapes of the light patches change in a quite dramatic way into little crescents. At the maximum phase, the patches can be seen as a distinct circle, just what we need.

The pinhole effect is best if the tree has small, well-spaced leaves and there is little or no wind. At the May 1994 eclipse the better sites were well into the desert regions in Arizona where trees are few and far between.

Fortunately, those trees were more like bushes of the acacia type with small and well separated leaves, not a dense overhang, absolutely perfect to show up the pinhole effect to advantage.

Some simple pinholes can be made with the obvious hole punched into a piece of card, and these work perfectly well. Other more elaborate devices can be made by allowing sunshine to pass through loose raffia work as in a straw hat. The most bizarre example I have come across is a hole punched in a dried cowpat! No eclipse view is ever really complete without some form of optical aid, and binoculars must be the most popular. Use any you already have, making sure that the filters are 100 per cent safe in terms of the density, but also that they are firmly attached and will not drop off when swung across the sky. Always spend a lot of time checking over these simple safety points.

What sort of binoculars to buy for the first time is never an easy question to answer. It is hardly worth while buying specially for a solar eclipse, even for regular solar observing; a simple telescope at a permanent site is going to prove more useful in the long-term. Choose some to use for whatever other hobby you have, or are likely to think of on the way to the eclipse. A general purpose, lightweight modern type is unlikely to be a disaster for these other uses.

In choosing a pair of binoculars, always insist on trying them before leaving the shop, preferably with someone you know who has some knowledge. Cost is no guarantee of quality or acceptance. Try slinging them round your neck for weight, and check how easy they are to grab in the dark (eyes closed) as you will be using them for night-time astronomy most of the time. If more than one in your family or party is going, try to afford one pair per person. Short eclipses are too valuable to waste time exchanging a single pair.

The final items to collect are good quality shoes and seats. There will be a lot of standing around and comfortable shoes made for the expected site conditions are a must. Portable seats are a desirable luxury, many would say an essential. Eclipses near the zenith, such as in Santorini in 1976, would also greatly benefit from a ground sheet or blanket.

A degree of eccentricity is allowable, and a shooting stick is not that offbeat. A wide-brimmed hat in hot climates adds the crowning touch. Sunstroke and heatstroke are very real hazards, and must also be included in your preparations for safe viewing.

Photography

All forms of photography from the simplest 'fun' camera to the latest electronic camcorder need practice and more practice.

Obviously, a 'fun' camera is not going to take long to master and will be used just as intended to collect lots of 'happysnaps'. I am not suggesting their use as a joke. They have a very serious part to play on every photographic expedition as a back-up. They invariably work perfectly (because they have never been used before, and are intended to be used by the fumblefisted!), and the optics can give some outstanding quality results.

Fill your pockets with as many as you can as a back-up, because they are ideal to take lots of pictures of the general scene around you and to capture those unexpected events and people when your main photographic equipment is locked up in your main experiment. These cameras cost little more than a reel of film, and so work out very cheap to run, most of them being recyclable. If you feel confident enough, recycle them yourself and use as slow a film as you can for even better quality results. One thing must be emphasized, do not use the flash (if fitted) during an eclipse, even an annular, as it will not only aggravate everyone else, but lose vital time regaining dark adaptation. Leave strictly alone all cameras working only with a flash, and do not bring them on to site.

To repeat: 'fun' cameras, particularly modern ones working as panoramics can be the best ones to use for general scenic shots, and often get pictures quite impossible with more complex alternatives. They are worth their weight in gold simply for the peace of mind they create.

Cost is usually the decider for the more complex equipment. However, simplicity is much more important. A totally new camera can be a recipe for disaster unless sufficient time and trouble is taken to get used to the controls, which must be second nature. This is true whatever type of recording medium you opt for. Some auto functions may suit you, but they must be checked first, particularly with video because of the sensitivity to infrared mentioned earlier.

Because it is rare to find a camera which is not black these days, remember that the thing is going to be left in the Sun before, during, and after the eclipse and the electronics are going to get very hot indeed. Select or make a cover, preferably a white one, to reflect heat.

Because of the superior image quality needed for the large image size involved, SLR cameras still remain the most popular choice for annular eclipses because of the wide range of lenses made. Make sure that you have enough adaptors and fittings when using lenses not designed by the camera maker. Check over battery consumption and buy only fresh batteries with plenty of spares, before you leave home. Battery consumption increases alarmingly in damp and very cold climates. This factor of low battery capacity at low temperatures destroyed many potential pictures at the total eclipse in Chile in 1994, simply because the batteries died after a few frames, or did not have enough energy left to wind on.

To recap the advice given in 1990, the history of eclipse trips is littered by tales of photographers failing to get pictures because they have not had a camera long enough to get used to its funny ways. Stick with an old one if in doubt. Better still, take your old camera along as well. All professional photographers carry a spare (or two!).

Today, you have 'fun' cameras as a further option.

Video workers have little option on the recording medium to use, it boils down to buying the best quality or make you can get. Film workers should adopt the same concept. There are not really any 'bad' makes of film these days, which means buying the slowest speed film you can get away with for the expected shutter speed, which in turn is determined by the solar filter. The range of shutter speeds is nowhere near as wide as for a total eclipse because there is no outer corona to worry about, although some marginal annular events will show a trace of inner corona.

Film workers should opt for a film speed of 100 ISO or lower to get the best resolution and contrast. As the image will be monochrome, there is considerable merit in using a black and white film only for Mylar filters. The images can be printed on to colour paper later in any colour to taste. The main snag with this is that the subtle gradation of colour towards the red/orange can be lost, but that is a small price to pay to get away from a hideous blue.

Optics

Sheer image size is not the only criterion – quality has to come into it. The longer the focal length chosen, the more care has to be taken to check the system as a whole for camera shake and stability. Also, will it fit comfortably into your luggage?

Large images are necessary, but it is also necessary to choose only up to the longest focal length conveniently working within your chosen film or video format: that is, the image size must be large enough to be useful but not so large that it is difficult to keep within the field of view. In practical terms this usually means a focal length around a metre (1000 mm) for 35 mm, which gives an image size close to 10 mm. This gives about half a solar diameter clearance either side across the width of the film, usually enough to compensate for any 'flop' in the mounting when brought to the field centre and released.

An equatorial motor drive makes life so much easier that one ought to be included in your list of essentials. However, not everyone can afford them, and there is also the problem of transport. A focal length up to a metre then comes into its own, as it is relatively easy to arrange the mounting so that the solar image drifts down the long length of a 35-mm frame or video screen. As a first approximation, this image drift takes two to three minutes whilst still keeping all the solar image in the field view. For many of the shorter events this is more than time enough to get all the pictures you need between second and third contact. It is certainly long enough for routine solar observing.

The time restraints are considerably eased if a larger format camera can be used. It takes a good five minutes for a metre focal-length image to drift far out from the centre of a 6×6-cm film frame, once started close to an edge. Do not use a zoom lens if you can avoid it, even with modern optics – always select the best 'prime' lens you can afford. Quite often a simple glass lens will outperform a zoom for reasons not relevant here. Suffice to say that the risk of 'ghost' and other secondary images and lowered contrast is too high. Mirror telelenses and telescopes are generally satisfactory.

Do not add tele-extenders until the particular combination has been tested fully. You will get a severe light loss and image degradation, usually manifested as much poorer contrast. By far the more serious effect is the worsening of camera shake.

There is a very important factor to consider at annular eclipses when using longer focal length lenses, particularly mirrors. This is the expansion of the optics as they warm up in the hour or so before second contact. There is little you can do about this except to try some form of cover, but that does little more than slow the process down as you are still getting a fair 'slug' of solar energy straight down the optics which are normally in closed tubes. Much better to be aware that it happens and check critical focus at regular intervals, more frequently as the important events get closer, such as sunspot occultations.

One useful dodge is to consider using a stop to make the focusing errors less critical. With some of the cheaper mirror optics, this has a dramatic effect on image contrast as well. Some say it almost makes the cheaper items usable! Be this as it may, it is worth recommending at all times with solar work because of this improvement in image focus latitude and contrast improvement, to say nothing of the heating reduction. The thing to use is an annulus, that is a ring with a hole smaller than the main optics entrance. If you are using one of the simpler camera mirror lenses, a stepping down ring does the trick very neatly at minimal cost.

The trick seems to work in two ways. First, it increases the effective focal ratio, hence the improvement in focus latitude, with part of the contrast improvement. The main improvement seems to be elimination of the very common edge/rim figure errors in the cheaper optics. With some of these in my possession, a terrestrial image borders on the just acceptable, but quite useless for solar imagery. The same optics become nicely focused when stepped down, a very acceptable financial saving over better quality optics. There is one final check to make on the optics, particularly if the focal length is long and you have difficulty using a camera viewfinder without some optical aid like glasses. Make test shots a little out of focus either way and see if it improves the image sharpness. If it does, make a note on the lens where ACTUAL focus occurs. Then check if there is a change each time you re-assemble the set-up. It will use up some film, but better now, and not when the exposures are for real.

Camera Shake

This phenomenon has been singled out as a separate discussion because it is by far the most important single cause of degraded images in solar photography with conventional cameras. It can happen with video, and for the same reason. Heavy lenses and attachments stuck on the front of the camera upset its natural balance and the slightest vibration in the mechanics start a 'nodding and rocking' vibration. Even with perfect mechanics, wind buffeting can, and usually does, play havoc.

A very sturdy tripod becomes the most essential ingredient in the whole system.

It cannot be emphasized enough, but it is true, that money spent on a good tripod or support is often a much better investment than a fancy lens.

Effectiveness is the only important criterion. I now make a simple but rigid structure customized for the anticipated elevation, and use that. When the eclipse is over, the cheap construction is left behind to make way for baggage allowance for souvenirs.

The tripod sturdiness must be checked for the longest focal-length lens and is best done by test exposures. But before you do that, tap the lens. Does the image take longer than 1/10 sec. to settle down? If so, the structure is unsuitably flimsy. Try the effect of adding weights or bracing struts. Also try the effect of a hairdrier at close range to simulate strong wind buffeting.

If the tripod is satisfactory, it is still possible to get camera shake because of the instability of the optics stuck on the front of the camera, when the only mounting point is the standard camera bush. Whenever camera shake is encountered, always check the natural point of balance of the system. If this is not where the camera is mounted on the tripod, you can forget about sharp images, unless you are very lucky, because of the natural vibrations about the incorrect point of balance. You must devote time and effort to finding some means of mounting the camera at its natural centre of balance, even if this means using a different optical system with a correct tripod bush.

Exposures in the most useful range of 1/125 to 1 sec. give most bother. Try some test exposures in this range. Do not put real film in the camera, just a dummy film in to get the balance right. If the front point of the lens jumps about after exposure, camera shake is a real problem.

The only thing to try now is another useful technique or concept of a decoupling damper. This means placing a thin layer of something flexibly rigid between the camera/lens system and the tripod mount. In everyday language, see if a thin pad of natural leather (or something with similar properties like cork) stops the vibration. It is amazing how such a simple pad can improve the sharpness of your images out of all recognition.

Having made sure the equipment will sit on the mount comfortably and without vibration, the final check is to see if you can line it up with the Sun at the anticipated altitude and azimuth.

A wide-angle lens obviously has no problem in lining up. The same is not true of the long telephotos and telescopes we hope to use. Lining up on the Sun with a heavy filter is surprisingly difficult, particularly in the haste often necessary on a crowded eclipse site. Time and effort designing some form of finder is well spent. It could be a back-up camera which is used as a counterweight at the same time. Check that you can consistently align whatever device you choose to your intended accuracy and that whatever it is that it does not snag vital parts at the anticipated elevation. Pay particular attention to balance and drive irregularities.

Suggested Observations

Work out a planned sequence of observations. Keep it simple and write it down. Work through this plan in dummy runs, giving at least 50 per cent of the time to just stand and stare. Working with a simple 'Walkman'-type player makes life a lot simpler, but remember that other people on the day might not want to hear it.

1 Always take plenty of pictures of the site as you arrive and set up. That is why you need those 'fun' cameras. Most eclipse viewers pack up immediately after totality and are never seen again!
2 If you only have one main camera, change film at the first opportunity to the right one, check the settings, then leave well alone.
3 Set up the equipment. Check that all is in order and do the necessary if there are problems. Use your audio prompt.

4 Take the absolute minimum of pictures during the partial phases, remembering that sunspot occultations are fascinating and you might want to change your mind. You will know if this aspect is a possibility before first contact.
5 Take lots of pictures with your back-up cameras of the pinhole effect and any other phenomena, such as birds going to roost or flowers closing.
6 One minute before second contact, double check that the camera is correctly set for Baily's Beads.
7 Judge the Baily's Bead stage, and take lots of pictures.
8 Take plenty of pictures during annularity, not forgetting to look around and at it, using filtered optical aid if taken. At annularities close to total, it might be possible to see some corona. This was certainly true in 1984.
9 Take more pictures at the second set of Baily's Beads at third contact.
10 Re-look at the eclipse, and general scenery.
11 Repeat instruction 4.
12 Relax or finish the film on pinhole effects or the scenery, wildlife, or whatever is seen.
13 Double check dismantling and packing procedures, autocueing if necessary.
14 Double check that films are correctly labelled and stowed away for the journey home.

This suggested plan is amongst the simplest, but, even then, has fourteen important steps to carry out. The need for practice beforehand should be abundantly obvious.

The Golden Rules of Eclipse Chasing

The Golden Rules of eclipse chasing boil down to five DON'Ts:

1 DON'T be too ambitious.
Take photographic equipment, but be prepared to get nothing from it and just use the 'Mark 1' eyeball, preferably with binoculars.
2 DON'T change an established system.
If you have tried and trusted equipment, stick with it. The eclipse site is not the time and place to find problems with the latest 'Gizmo' with flashing lights.

3 DON'T panic.
  No explanation needed. See Rule 1 (above).
4 DON'T lend anything to anyone.
  Lend equipment after the eclipse by all means. Before that, Jack really is Number One.
5 DON'T let your film and pictures out of your custody.
  Your pictures are a unique record and memento. It is as near to a certainty as anything can be in this world that if you lend original pictures, they will either be lost or be returned damaged. Make copies specially for loan or reproduction.

Some Special Annular Eclipses

All annular eclipses are of interest and some scientific value. The really nice ones are those very close to a total and it become a close-run thing whether or not the corona can be seen. In many such cases some chromosphere is visible, and prominences spotted.

Until very recently there was a heated discussion in astronomical circles whether or not the Sun was shrinking or even expanding. The idea here was to check both theories with observations at annular eclipses, particularly close totals. Some estimate of the Sun's size could be gauged by how far out the track was from prediction, and this was best done from records made in the historical past. The evidence seemed to be somewhat ambivalent, and great effort was devoted to measuring the solar image from Baily's Beads, assuming a mean diameter for the Moon.

That matter now seems to be settled and we can just enjoy the spectacle and not worry too much about the solar size.

1976, April

This annular was a particularly interesting one as it occurred close to local noon, almost at the zenith on the fascinating Island of Santorini. It was local spring with an abundance of large daisies in bloom. These started to close as the eclipse progressed. Close to second contact birds were seen going to roost.

Annularity was a particularly long one, around six minutes, but not all that spectacular visually. Nevertheless, the sky darkening was sufficiently deep to fool cockerels which crowed shortly after.

It was the setting making the eclipse memorable, particularly with the island's association with the Atlantis legend, and the central volcano, responsible for so much devastation to the Minoan civilization. That civilization is probably responsible for much of the astronomical naming and cataloguing handed down to us via the Arabic period.

1984, May

The main claim to fame for this one is the huge number of people seeing it from USA. It was also a close total and many rare (up to that time) recordings of the chromosphere and inner corona were reported. The weather was somewhat variable, but that did not matter because the eclipse also set another trend in being widely covered by live TV.

Some observers were able to see the event live, then have the delight of an 'instant replay' from another site further down the track. In this way some people clocked up many minutes of 'live' eclipse viewing.

1994, May

This was a Saros on from the Santorini annular of 1976, and of similar long length, around six minutes. It was my second Saros, completing a 'set' with the total from 1973 and 1991.

It was another one seen by a large number of people in USA, cutting across much of the middle, starting from Arizona and Texas. As this is largely desert, the conditions were generally excellent although some unseasonal cloud and rain was encountered.

As noted in the text, the 'pinhole' effects were very pronounced because of the leaf structure of the desert plant life. It was also a nice easy one to observe with the Sun rising almost straight up at that time of the year.

1995, April

An annular eclipse took place almost a year later in April, cutting across South America almost along the same track as the total a few months earlier. There will not be any annular or total solar eclipses in 1996, but there are a couple of nice partials to practise on.

1996

The first is to be seen from New Zealand on April 17. The maximum phase will be 88 per cent over the southern Pacific Ocean, which means that many of the biological phenomena with plants and animals might be noted on land. The second partial eclipse takes place on October 12. It is visible from a large number of densely populated areas, including part of Canada, Europe, and Northern Africa. The more adventurous might like to consider Greenland and Iceland. The maximum phase is just over 75 per cent, allowing plenty of scope for sunspot occultation studies as solar maximum is close.

1997

The next eclipse is a total on March 9, visible in Mongolia and Siberia.

1998, February and August

Another total eclipse on February 26 is only visible over a little land in islands and the top of South America.

It is not until August 22 that the first new annular eclipse takes place. It will be seen from Sumatra, passing through Malaysia and Borneo and some small islands. Perhaps the best views will be seen from the sea.

Annularity is a maximum of nearly 2.25 minutes. Close to sunspot maximum this eclipse could prove to be interesting.

1999, February

Only a few months after the last annular, on February 16, this one is relatively short at a little over 1.3 minutes maximum. Around a minute of annularity should be visible over a large track through Australia.

2001, December

This is a poor one to see over land as nearly all the track is over water. The best place to be on December 14 is in Costa Rica or the south-east of Nicaragua. Over the sea, the maximum is nearly 4 minutes.

2003, May

This is a weird event on May 31, having an altitude of 3° or less. As it occurs over the northern Atlantic Ocean, the chances of seeing it from land are limited to parts of Shetland, the Faroes, Iceland, and Greenland. The furthest southern landfall is the tip of Scotland.

Because of the geometry at that time of year, the eclipse has no northern limit. The strangeness of it makes this a trip not to be missed, even if the chances of clear skies are low during the nearly four minutes it takes place.

2005, April

Only a couple of years later another rare event occurs on April 8. Here the track starts off as an annular close to New Zealand. By the time it has reached the Oeno Atoll it should switch to being a total for a few seconds, probably still being a total still over the Galapagos Islands. However, by the time it reaches land again in Panama it will be back to an annular event. This is one eclipse not to miss. The Baily's Bead effect is dramatic enough during any annular eclipse, but for this one the 'String of Pearls', seen as the two limbs almost precisely match, should be at their very best. Save up and go to this one if you can.

Starting with a CCD Imaging Camera

KATHLEEN OLLERENSHAW

This is an account of how and why I came to acquire a telescope so late in life; of how circumstances led me to acquire a CCD (charged coupled device) imaging camera, and of the boon which this has proved to be, bringing me great joy.

Late one evening in the summer of 1990 I was driving down the M6 from Scotland to my home in Manchester. Above was a remarkable and unusually clear star-studded sky. Suddenly it occurred to me that, with the commitments of a packed academic and public life gradually lessening as the years passed, I could, if I wished, satisfy a long-held desire to buy a telescope. I contacted, and joined, the Manchester Astronomical Society: a very happy decision. An adviser suggested an 8-inch Celestron Schmidt–Cassegrain and, within a week, this was set up on the small first-floor balcony at home. My first view was of a spider building his web on a neighbour's television aerial! I quickly achieved my initial ambition to see Saturn and its rings. With the help of the then president of the MAS, Kevin Kilburn, I had soon made satisfying prime-focus pictures of the Moon.

I did not know at that stage what a treasure trove of glistening colourful jewels there are in the sky, available for any of us to see – given a good pair of binoculars or a modest telescope. And I knew nothing of the pure pleasure to be had in picking out the constellations for the first time and seeing the changing aspects of the Moon in detail, not to mention the happy hours spent reading the marvellous mainly-new-to-me books on astronomy, its history, and its practice. With hindsight I now firmly believe that the excitement of exploring the night skies for the first time is perhaps even greater for someone no longer young than for those who have had this enjoyment since their youth. Given the right background, it seems never too late to start.

Although a complete beginner in astronomy as such, I did, to be fair, have some advantages. I am a (pure) mathematician by profession. My late husband, a radiologist, was an accomplished photographer who was always needing an assistant (me!), and I had a good camera. More importantly, as it transpired, I had been using computers since punch-card days. My original interest in computers was aroused immediately postwar. I am an exact contemporary of Alan Turing who, after the war, joined the mathematics department at Manchester University where I was working as a part-time lecturer. By 1990 I owned a 286 portable Compaq computer. Like most of us, I had followed the space programmes during the 60s and 70s with intense interest – indeed I shook hands with Yuri Gagarin when he came to Manchester a few weeks after encircling the earth in 1961 – and I had gone to some lengths to try to see Halley's Comet in 1986 from an aircraft; but I knew next to nothing of the constellations and stars themselves. If I was even to get started I would need a lot of help, and in this I have been extremely fortunate. I got to know a great many young people willing to give time and effort and share their hard-won expertise with someone much older than themselves – because they love their subject so much.

There is no substitute, given a clear night, for enjoying the night sky with the naked eye, assisted maybe by binoculars. But, to quote Patrick Moore's admirable tape *The Sky at Night*, 'the stars become so much more interesting when you know which is which' and I wanted to learn quickly. Eyesight deteriorates as the years go by, and I soon realized that I was unlikely to be able to sustain an active interest in this new activity if I limited myself to observing only by eye. By far the best way for me to learn on my own, on the few and unpredictably-occurring clear nights vouchsafed us in the north, seemed to me to be to take as many pictures as possible, when the opportunity arose, and then compare them with the maps. Even the most magnificent of star atlases means very little unless related to what one has actually seen. The thrill of matching asterisms in one's own pictures with those in the maps came as a surprise.

Under Kevin Kilburn's guidance, I set about mapping our own area of sky with 3-minute exposures on fast film, using a 35-mm second-hand Pentax screw-fitting camera, piggy-backed on top of the telescope. In order to take advantage of darker skies away from city lights, I also established a base in the country to which I can go whenever I am free. The roof of a disused earth closet, $4' \times 4'$ in cross section, was replaced by a sliding wooden top. This first 'observatory' was duly opened with a champagne party and christened (with Sir Bernard Lovell's permission) Lovell II.

In the event, Lovell II proved too small for its purpose and my first pictures were taken standing in the open, where, in due course, a steel pillar, 6″×6″ in section, was sunk deep into the ground and surmounted by an equatorial wedge to give fixed polar alignment. This, however, had the serious drawback that I had to lift the heavy telescope off the wedge and carry it to the shelter of Lovell II, in the dark, whenever storm clouds threatened. Sooner or later, either I or the telescope, or both would have a nasty accident. Moreover, transporting the telescope to and from Manchester in the boot of the car held no long-term future: something more permanent was essential.

Meanwhile, in July 1991, I went to Hawaii and was one of the lucky members of Explorers Tours who had a clear glimpse of the eclipse. When Patrick Moore used my picture on his BBC *Sky at Night*, this inspired me to even greater enthusiasm. The next priority was therefore to build a proper observatory.

As the photographs (Figures 1 and 2) show, this observatory is built to take advantage of the slope of the ground. A couple of steps lead up on to an 8′×8′ interior floor. Breeze-block walls, wood-panelled to shoulder height, precisely levelled, with runners set in concrete along the top, go all the way back to a high stone wall

Figure 1. The Observatory, front view, 1994.

Figure 2. Side view of the Observatory showing construction to take advantage of ground slope.

which forms the northern boundary of the backyard. On the runners rests a heavy wooden superstructure, 8′×8′ in cross section, 4′ high at the front, 2′ high at the back, with a sloping felt-covered roof. This top is fronted by two double-hinged 'stable-type' doors which fold outwards and are hooked back to its wooden sides. With these opened and secured, the whole top slides back along the two extended breeze-block walls of the base to leave the 8′ square inner space completely open to the skies. The mechanism is provided by a second-hand bicycle chain and gears, the top being wound back and forth by means of an old car starting-handle fixed on the outer side of one of the walls of the base. All this was achieved at minimum cost and in minimum time with the help of a neighbour. The new observatory faces due south. There are trees to the east and hills block the western horizon below about 30°. The high wall into which the winding mechanism of the sliding top is fixed limits the northern aspect to about 20° south of Polaris, which twinkles through trees above this wall. The position is excellent and there are no intrusive artificial lights.

The heavens were not willing to be spied on without a warning. Just as the new observatory was nearing completion, a small tornado swirled in over the hills to the west and lifted the 2-cwt top clean off the runners, hurling it sideways for some ten feet against the stone wall. I was watching the fearsome storm from the safety of a bedroom window. Anyone out in the backyard and near the observatory could have been instantly killed. After that, extra bolts were fitted to secure the top when closed.

One Easter evening with exceptionally good skies, while on holiday near Bowness, I visited Sally Beaumont. She has for many years been an active member of the British Astronomical Association with a particular interest in the planets, and was awarded the BAA Messier Certificate for observing and recording all the 103 Messier objects which can be seen in northern UK skies – this with an old Browning 3-inch refractor and no digital setting circles! She found and showed me a number of Messier and other deep-sky objects which I had until then only seen illustrated in books. This further whetted my appetite – it had to be the deep sky. But finding the desired faint deep-sky objects and centring them in the telescope was going to be a problem. Moreover, ordinary photography with the long exposures needed was likely to be impractical. I no longer have darkroom facilities, clear skies are rarely reliable even when they occur, nor were the difficulties and discomforts of guiding for any length of time an attractive proposition; necks, shoulders, and backs become less accommodating as one gets older. While I was pondering on this, Denis Buczynski invited me to his observatory near Lancaster. He is also an active member of the BAA with a special interest in astrophotography and the accurate measurements needed when searching for and photographing comets. He showed me an image of the Ring Nebula obtained with his newly acquired Starlight Xpress CCD camera. This, I felt, was the answer. If I had ambitions to obtain images of deep-sky objects for myself, it would have to be with a CCD.

So far I had not known in what direction I wished to go. Now that my objectives were becoming plain, I had to consider up-grading my equipment. Circumstances largely determined the choices. There was always the feeling of being in a hurry, mainly because of advancing age, but also I had come to realize how scarce are the clear, moon-free nights which coincide with occasions when I can get to the observatory. The solution was to invest in a telescope capable of finding required objects for me which could track reliably for up to five minutes or so, and to acquire a suitable CCD camera, preferably designed and developed in the UK, so that advice during the learning phase would be available: indeed, Terry Platt's Starlight Xpress.

I had earlier acquired a Santa Barbara Instruments ST4 CCD camera with the idea of countering the bright street lights at home in Manchester. Excellent though it may be as a tracker, as an imaging camera I found it somewhat disappointing. For the country, I began to prepare for a Starlight Xpress, and indulged in a Meade LX200 10-inch which would find the deep-sky objects on command. This decision was reached somewhat reluctantly, but common sense ruled. When the rare opportunity to get good pictures presents itself, the best use has to be made of every moment; time spent in manual searching would bring insufficient returns. I was much helped by the staff of the suppliers, 'Scope City in Liverpool and Broadhurst, Clarkson and Fuller Ltd of London EC1, who take a personal interest, far beyond commercial advantage, in answering a beginner's naïve questions, and who have twice made the journey north to solve specific problems. Without, in addition, the continuing practical advice and working visits from Manchester of members of the MAS, I might have become hopelessly frustrated by minor problems. In particular, precise polar alignment is a *sine qua non* of accurate operation of the programmed telescope, although many advertisements, emphasizing portability, seem to treat this lightly. I was specially grateful to MAS friends for helping me to improve alignment – no easy task. The greater the accuracy, the better the tracking, and the more chance there is of arriving on target after slewing. Even a fraction of a degree in increased accuracy makes a world of difference and can double the possible length of exposures without visual guiding.

When the observatory was ready and the new telescope in place, Terry Platt drove up from Berkshire, and installed a complete Starlight Xpress together with a secondhand IBM 286 computer and an old redundant monitor in order to get me started. His practical help then and subsequently has been of inestimable value.

For exploring the broader sky, I was advised early on to buy an f/6.3 reducer which, with the increased field of view, makes finding particular stars much easier than when working at f/10. I now also have a long-focus 300-mm lens mounted permanently on the telescope for semi wide-field photography. There is a home-made fitting at the back into which the CCD camera can be firmly pushed to give pre-set focus which needs at most only slight adjustment before each take. The CCD camera can also be linked to the 35-mm screwfitting lens of the Pentax camera to take wider-field images. This gives a satisfying range of options available for different deep-sky objects. All these additions gave me trouble at first when trying to arrive at the correct focus essential for success, but the potential, given enough opportunity, is tremendous.

There were and still are many tough lessons to learn. Any earlier experience, however sparse, with a telescope or in photography is an advantage. Computer literacy, which will soon be as common as competence on a typewriter, is also useful. Although it just about sufficed, I found the 286 computer far too slow to be tolerable for long. The faster the computer and the more hard-disc space available, the better. The make-do monitor had also to be replaced with a recommended 9″×9″ standard surveyance model: unless the quality is sufficiently good, the 'noise' completely obscures most deep-sky objects, so that finding a faint object becomes virtually impossible.

The first task, having located the desired target in the telescope as best one can, is to centre it on the tiny chip. I failed to do this with any certainty until I acquired a dual-position hinged-mirror diagonal into which the CCD camera slots for the usual view. A favourite 20-mm Plossl eyepiece was then brought into exactly matching focus with the CCD camera and a collar fixed around its stem so that the eyepiece and the camera are now always in focus simultaneously. This means dedicating the eyepiece to this one purpose. The Plossl has no cross hairs so that initial centring of an object on the chip, and hence on the screen, is achieved only by guessing, but its wide field of view gives me the considerable advantage of being able to observe the sky, wherever I choose to point the telescope, without having to disturb anything. An illuminated reticle eyepiece is best for fine alignment and for checking the accuracy of the hinged mirror.

A difficulty which I had already experienced with an ordinary camera body attached directly to the back of the telescope, was that

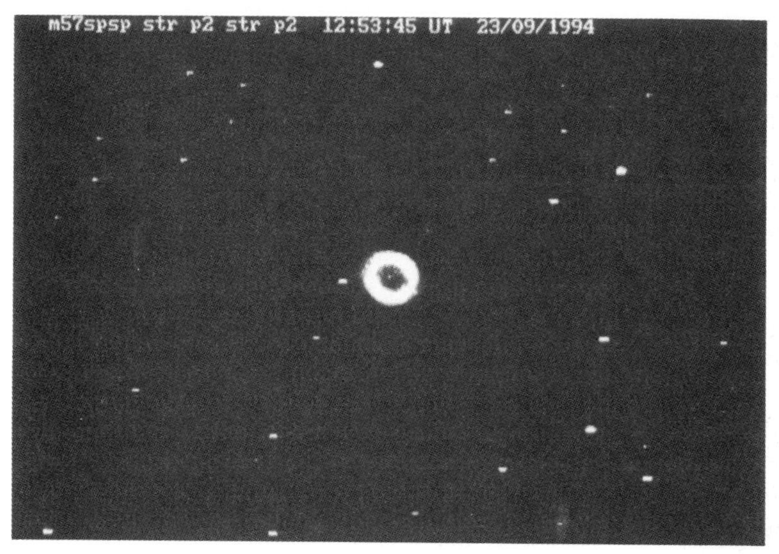

Figure 3. M.57: computer printout of CCD output.

Figure 4. M.13: computer printout of CCD output.

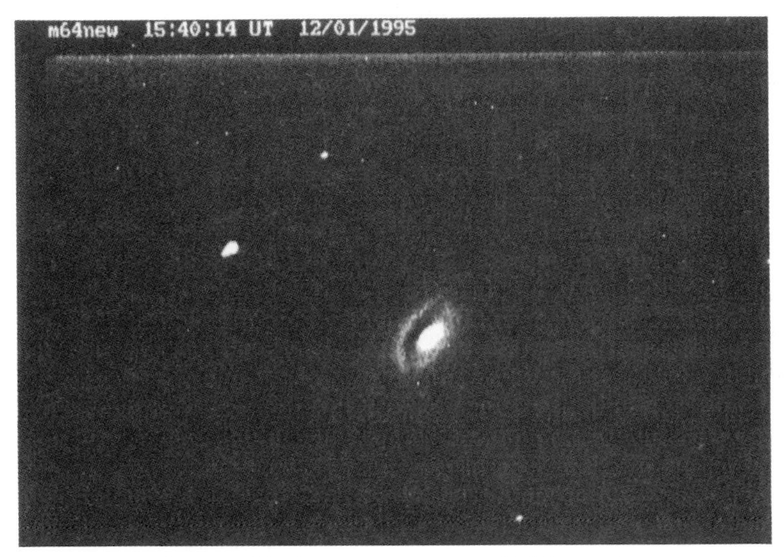

Figure 5. M.64: computer printout of CCD output.

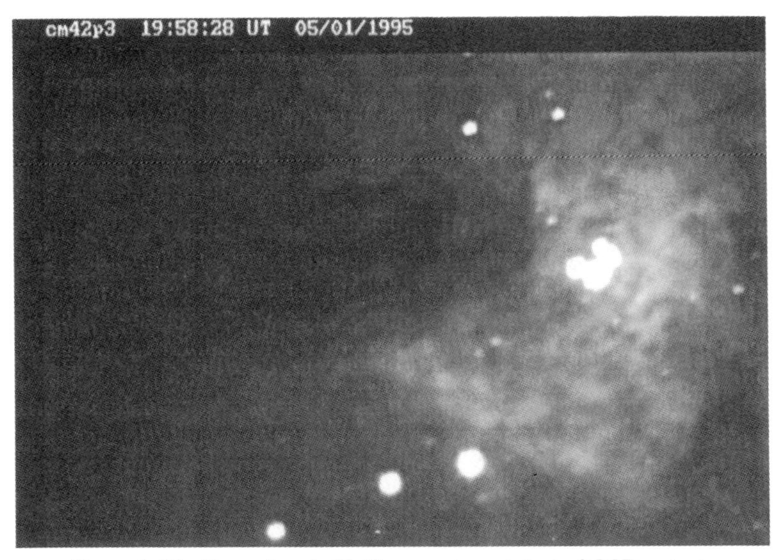

Figure 6. The Trapezium in M.42: computer printout of CCD output.

my eyes focus differently, whereas film and the CCD chip lie in their own respective fixed planes and are unaccommodating about focus. (I now search with one eye and focus with the other, but even this does not necessarily bring accuracy!) With focusing on film this was very unreliable. Moroever, it meant waiting to finish a 36-frame roll and then a further time gap for professional processing. At best this is hit or miss, and re-tries have to wait for the next clear night. One of the great joys of using a CCD is that this fundamental problem is solved at a stroke. Best focus can be obtained then and there on the screen. There is instant evidence of success or failure and opportunity for as many repeats as desired.

Even the most sophisticated modern telescope needs its points of reference. When the telescope is permanently housed and accurately polar aligned, with Polaris then automatically registered in the telescope, it is sufficient to bring into view and register just one other known star. When first setting up for a viewing session, any conveniently placed and known bright star is chosen and its code-name punched into the telescope's hand control box. The telescope then slews to approximately the correct position. Because it has been closed down since the previous session, it may at this stage be a considerable way off target. The star is then centred manually, with or without the help of the digital declination setting circle and the telescope's direction buttons. When the chosen bright star is then brought into best-guess focus and centred as accurately as possible in the eyepiece for the purpose of initial registering, the hinged mirror is switched over and the image of this star will, with some luck and a little experience, come up on the screen. With the exposure cut to a minimum to save time, it can then be precisely centred. This image may be very bright and it is a good idea to fix a thick, red cellophane sheet to the top of the monitor which can be brought over to shield its screen. The star is then registered in the telescope's memory (on the LX200 this is done by pressing 'co-ordinates match' on the hand control), which, from then on, will, on command, cause the telescope to slew to any desired target whose co-ordinates are stored within the telescope's programme, or to any other desired position above the horizon, the known co-ordinates of which are 'punched in'.

Once the position of this chosen star has been registered, the telescope can be slewed slightly to bring smaller stars on to the screen. The focus can be further refined, actually on the screen itself where it matters and using *both* eyes together (with specs if need be). All this is done with exposures of not more than three to four seconds. The advantages of the process for someone in a hurry, in comparison with using film, are immense.

The targeted object has now to be found. If this is not close to the star chosen earlier as a fixer, time is well spent in slewing the telescope to another known star nearby and checking again the accuracy of the centring on the screen. With a faint object in a dark patch of sky, no image will appear at all unless there is adequate exposure. The exposure time thus has to be increased to at least a full minute, and maybe for as long as five minutes, for even a trace of the object to be discernible. With good focus already assured by testing on a nearby, but not-too-bright star, and an adequate exposure time allowed, the programmed direction-finding of the telescope should then ensure that the desired image is at least faintly visible. With the image anywhere on the screen, it can be centred by using the telescope's slewing controls at the slowest available setting, and the focus refined yet again.

The next step is to take the picture. The exposure time, which one would like to be as long as possible, has to be limited by the accuracy and reliability of the telescope's drive, or of the hand or other form of guiding. The CCD chip is so small that even the most minute accidental movement or atmospheric wobble results in trail or double-imaging. However, even with generous exposure, several shots can be taken within a reasonably short time and only the best images used. When a satisfactory image appcars, it is saved into the computer, ready for processing later. This saving need present no special difficulties if the instructions are properly followed.

It seems appropriate here to tell of one of those silly errors that can plague a beginner. With film, the button on the camera's shutter extension cable is pressed to start a bulb exposure. At the end of the allotted time, the same button is pressed again to close the shutter. With the ST4 the technique when an image is on the screen is to 'grab', that is, to leap to the computer and start pressing keys before the image moves away. With the Starlight Xpress software, procedures are, happily, much less fraught. The dials on the 'frame store' having been correctly set, the required image centred and in focus on the screen, the exposure is then started by pressing a button which is not unlike that on an ordinary camera's shutter extension cable. At the end of the exposure time a new image flashes on to the screen *and remains there*, static and fixed, awaiting the next move – either to save on the computer or to retake. At first, when there was a good image, habit made me hurriedly press the button again as is done to release the remote-control camera shutter after a bulb exposure, but with Starlight Xpress this is the procedure for starting afresh. A new exposure would thus be initiated and nothing was securely saved. I was distraught because of this repeated and inexplicable failure, wasting unnecessarily a couple of rare clear nights – and all because I had not read the instructions carefully enough!

An unexpected bonus of the CCD camera over film is that good deep-sky images can be obtained even when the Moon is up, because of the short exposures required and the narrow field of view, to an extent not possible with film. With a cloudless sky, when taking CCD pictures of the Moon – the best target of all for a beginner as it is always easy to find, some method is needed to reduce the intake of light. I was lucky at my first attempt of imaging the Moon's surface, a few nights after the CCD camera was installed, in that there was a thickish mist which provided exactly the required filtering. A simple method I had to hand thereafter was to use a proprietary 'Kwik Focus' telescope lens cap, which has two 2-inch diameter holes useful for obtaining accurate focusing for the Moon and planets. Using one hole fitted with a high-density filter and blocking the other gives a quick and easy method of arriving at acceptable Moon images. The real solution for serious imaging of the Moon and planets is, I understand, to increase the magnification by using a suitable Barlow lens or similar technique which increases the size of the image and thus brings the correct exposure time to within a manageable range of options. I have not yet had the opportunity to try this and regrettably had to be content with other people's pictures of Comet Shoemaker–Levy's impact on Jupiter in July 1994.

One useful practice I have adopted is to establish a new computer file at the start of each session in the observatory – using the date, 950101 for example, as the file name. There are only seven free digits for naming the images, and variations for distinguishing them soon run out when some of the time during each session is spent trying to improve on earlier images of the same object. One of the special advantages of the computer is that it is no longer necessary to take notes at the telescope to record dates and time, length of exposure, settings used, and the object's celestial co-ordinates – all these can appear automatically in a computer heading. But, if sanity is to be retained, a simple filing technique is needed with the object's name and perhaps some further coding about the order of the take. If new images of the same object are placed in dated files, there is no need to invent different reference names. Eventually a decision can be made about which image of a chosen object is the best, and the others wiped out to save disc space.

Another useful trick concerns the dark frame. A dark frame is taken with all light excluded from the chip, thus recording the intrusive and unwanted electronic noise inherent in the particular CCD camera. This dark frame is then subtracted from the image taken of the object itself as the first stage in the retrospective image processing. The dark frame is related to the temperature and the settings used to control the exposure. Because it concerns only the electronics and not the optics, it is independent of where the telescope is pointing or of telescope movement. Dark frames can thus be made either at the start or during an observing session.

When a particular image is especially promising, it is a good idea to make and save a new dark frame immediately with exactly the same settings and exposure time. The obvious method of making a dark frame in the middle of an observing session seemed to me at first to be the same as when setting up, namely to put the storage cap over the telescope lens. But this can be a nuisance if the telescope is angled awkwardly or if a dew cap is in place; it is also ridiculously easy to forget to remove the cap and thus waste the next planned exposure time watching a blank monitor screen, awaiting the new image which never appears! A much simpler method when using a hinged-mirror diagonal is merely to switch the mirror so that nothing gets through to the chip – with the added advantage of then being able to scan the sky pleasurably in the eyepiece during the dark-frame take.

There is a slight problem which I have not yet learnt how to overcome. If alone, and therefore having to keep check on the monitor screen which is necessarily reasonably bright, as well as saving successful images into the computer, it is difficult for the eyes to become fully dark-adapted. The best conditions for viewing the sky direct thus do not develop. To compensate, when nearing the end of any session in the observatory, I try to give myself fifteen minutes spent gazing at the sky unimpeded by any telescope – if, that is, it is still clear.

With the CCD images saved into the computer and a dark frame made, there is still the image processing to be tackled before the results can be exhibited. The images may look really good on the small sharp screen of the monitor and yet be almost invisible on an ordinary computer screen. Image processing uses computer programmes designed to improve contrast, increasing or decreasing brightness selectively, and to extract information from the data which is recorded by the CCD camera even if not apparent visually. This is comparatively straightforward, but it requires good software, good instructions, practice, judgement and, above all, time and patience. Even with a fast computer, two or three hours can disappear in a flash while trying to get the best results. But, unlike developing and printing film, it can at least be done sitting down comfortably and in a warm room. The more powerful and faster the computer the better. With a slow computer (anything less than a 486 processor) it can become intolerably tedious as there is no real substitute, anyway while experience is building up, for trial and error.

There is then the matter of producing copies of selected images. Copies can easily be made of computer discs – indeed they must be regularly made as back-ups for chosen images – but not everyone has the facilities to display others' discs. Projection slides can now be made by commercial firms direct from disc, but this is expensive. However, the computer screen can easily be photographed on to slide or print film, and these photographic copies can again be copied to make enlargements of the original. These procedures are all within the scope of the amateur and without a darkroom, but time (and patience) is not unlimited. The simplest practical means of having something in the briefcase to show around is probably to use a good printer and learn the necessary computer graphics to make paper copies from disc. Familiarity with Microsoft Windows is becoming a basic requirement and I am currently teaching myself to use this powerful software.

It should be apparent from the foregoing that the most important factor for the success of the late-starting beginner is friends willing and able to help and advise, either in the flesh or at the end of a telephone line. I have been incredibly fortunate in this respect. I am continually amazed at the generosity of mind and spirit of so many amateur astronomers as well as those in the trade. There is a remarkable community spirit among amateur astronomers and pride in each other's achievements beyond anything I have met in other activities, sports, or hobbies. Perhaps this is because astronomy is still thought of as somewhat esoteric, involving voluntary long, cold, night vigils. It is not *competitive*, and is often (perhaps most often) carried out alone and desiredly so.

Fortunately, for astrophotography and for certain uses with CCD imaging, old cameras and lenses are usually more suitable than the modern electronically controlled and battery-dependent models. Film, both monochrome and colour, becomes faster and of finer grain by the year; high-powered computers and printers are rapidly falling in price, and telescopes properly cared for can last a lifetime.

There are hazards in every developing new technology, and sometimes it may seem as though there is an infinity of minor as well as major technicalities which can go wrong, each of which in turn can ruin a rare clear night. But when, after a few months, the worst frustrations are overcome and suddenly there is a magically clear, dark sky with everything going well, then what can compare with the elation of finding on the CCD monitor the myriad stars of, say, the Hercules Cluster in all their glory. This may be a travesty of the hard way of old, but a reward for the brilliance of the young generation of scientists who have developed these marvels of technology is surely that we should benefit by them, and use for our enjoyment what they have made possible.

Advances in this CCD field are so rapid and spectacular that any new technology is bound to be soon outdated, but the learning process is probably the same whatever the actual stage of development. Without any doubt my next endeavour will be to get into 'single-shot' CCD colour. It is already available and affordable.

It seems a miracle that I, a raw newcomer who, only a short time ago, scarcely knew Orion or the Plough, can today bring into view at will the Andromeda Galaxy, the Ring and Crab Nebulæ, the spiral galaxies M51 and M101, and the glorious globular clusters . . . and tomorrow show them to friends! Even if the images are crude compared with the magnificence of the large-scale colour photographs of the professionals, there is nothing to compare with the wonder, pride, and delight of producing an image all-by-oneself in one's own backyard.

Meteoroids from Interstellar Space?

DUNCAN STEEL

Whether or not particles arrive in the inner Solar System from interstellar space has long been a contentious issue. In fact Shakespeare seems to have anticipated the scientific debates on this topic when he wrote, in *Twelfth Night*, 'Out, hyperbolical fiend! How vexest thou this man!' – any object coming from interstellar space would be on a hyperbolic orbit.

If some object – a comet or a meteoroid, say – could be identified as coming from interstellar space, then it would be an important scientific result, because it would indicate to us that we could, in principle, sample matter from outside the Solar System. Such matter might, for example, pre-date all extraterrestrial samples that we have to hand, these having a maximal age of 4,550 million years from radioactive dating, that being the time since the solid matter in our planetary system – and, presumably, the Sun – agglomerated from the pre-solar nebula. To see how some object might be identified as being interstellar in origin, we first have to consider the different types of heliocentric (Sun-centred) orbits that can occur.

Objects in the Solar System under the gravitational influence of the Sun occupy orbits known as conic sections. The shapes of such orbits are described by the parameter known as the *eccentricity*, symbol e. A circular orbit has $e = 0$. All objects bound to the Sun, so that they cannot escape from its domain, must have $e < 1$, and such orbits are termed *elliptical*. All of the planets have low-eccentricity (meaning near-circular) orbits, Mercury and Pluto having the highest eccentricities, $e = 0.21$ and 0.25 respectively. Most of the asteroids in the main belt have low values of e, too, although some Earth-crossing asteroids have eccentricities as high as 0.9. Whilst there are a few comets with low eccentricities, most have larger values, like Comet Halley with $e = 0.967$. Such orbits are highly elongated. Comet Halley returns once every 76 years, but higher eccentricity orbits tend to take longer to return – the so-called *long-period comets*. In fact, some are not periodic at all, since if $e = 1$ the comet would have just enough energy to escape from the solar gravitational grasp, its orbit then being *parabolic*; and if $e > 1$ then the orbit would be *hyperbolic*. Ellipses, parabolas, and hyperbolas, then, are the three broad types of orbits.

Many long-period comets have been seen over history, and in recent times ten or more have been found each year, mankind having a fleeting chance to see them before they disappear into the depths of space, most likely never to be seen again. There are exceptions – for example in late 1993 my colleagues Rob McNaught and Ken Russell discovered a comet which was returning for the first time since the ancient Chinese recorded it in AD 574, setting a record for an observed return – but in general long-period (or *near-parabolic*) comets are only witnessed once. We believe that these have been thrown out of the Oort cloud, at a distance of 10,000–100,000 astronomical units from the Sun, and mostly blaze through the planetary region once before passing beyond our gaze for an interminable length of time. Passing stars and giant molecular clouds are thought to be the main agents causing Oort cloud disturbances. Since long-period comets have eccentricities very close to unity, their orbits are unstable in the sense of being bound to the Sun. It is like a drunken man walking along a cliff-top – one step in the wrong direction and all is lost. For example, if a near-parabolic comet passes just behind a planet in its orbit about the Sun, it will receive a slight acceleration by the planet's gravity, and thus be boosted on to a hyperbolic orbit, being ejected from the Solar System, never to return. Conversely, passing in front of the planet will tend to slow the comet down slightly, reducing its eccentricity so that it is more stable against ejection. One large (or many small) perturbations may eventually slow down a comet such that it enters a shorter period orbit (like Comets Halley, Swift–Tuttle, Grigg–Skjellerup, or Kopff).

Even though the Oort cloud is far from the Sun, the comets orbiting there – and perhaps occasionally being diverted into orbits passing close enough to the Sun for us to detect them – are all bound to the Sun and are therefore members of the Solar System. The question is, have we ever seen one which was *not* part of the Solar System? The fact that we know some comets to be thrown out of the Solar System into interstellar space raises the possibility that other stars might do likewise, producing interstellar comets that eventually make their way to our neck of the Galaxy. Equally well, one might expect some comets to be left as debris from the formation of other stellar/planetary systems. Do any of the observed comets appear to have come from interstellar space?

The answer to this is tied up with the intricacies of determining orbits accurately. Although a few comets have been observed to have eccentricities *just* larger than one, anyone looking for interstellar comets has to take into account a number of effects. First, it is not appropriate to use the centre of the Sun as the focal point of the cometary orbits; one has to use the *barycentre* (centre of mass) of the Solar System. This is shifted from the Sun's centre by the masses of the planets, Jupiter in particular. In general the barycentre is outside of the Sun itself, its location varying as the planets move in their orbits. When one calculates the orbits relative to the barycentre of potential interstellar comets, many of them are found not to be hyperbolic at all.

Second, there is the non-gravitational force applied to a comet by the jetting effect of the volatile material evaporating from it as it is subjected to heating by the Sun. It is this effect that leads to Comet Halley being four days 'late' in each apparition, at least over the past few millennia. For some comets the force acts in the opposite sense, causing them to return slightly 'early'; it all depends upon the mass loss rate of the comet in question, the distribution of the active zones of evaporation across its surface (like the vents on Comet Halley seen from the *Giotto* spacecraft), and the way in which it is spinning or tumbling in its orbit. Earlier I mentioned that with enough energy a comet can escape the Sun's gravity into interstellar space. That energy can be quantified in terms of its speed. A circular orbit at one astronomical unit from the Sun renders a speed of 29.8 kilometres per second; this is the Earth's mean speed (which varies slightly during the year since the Earth's eccentricity is 0.0167). Further out in the Solar System, circular orbits render slower speeds, whilst Venus travels a bit faster than the Earth. Now, if a parabolic comet passed by the Earth then its speed would be 42.1 kilometres per second, because it is a law of celestial mechanics that an orbit with $e = 1$ has a speed which is the square root of two times the speed of an orbit with $e = 0$. This speed of 42.1 kilometres per second is the 'parabolic limit' at one astronomical unit from the Sun. A comet moving at this speed as it passes the Earth's orbit might have its non-gravitational force applying a slight acceleration (due to volatiles – especially water – being lost as the solar heating becomes intense) and causing the speed to increase to 42.2 kilometres per second, and then the comet would be boosted into a hyperbolic orbit, never to return. Conversely, a decelerative non-gravitational force would slow the comet to 42.0 kilometres per second, making it safer against ejection. Recall the idea of the drunken man walking along the cliff top – this would be equivalent to him taking a step away from the precipice.

Of all the observed comets, the largest observed speed excess over the parabolic limit is 0.8 kilometres per second, and it seems that this can be explained as being due to the effects discussed above; that is, none of the observed comets can be shown to have come from interstellar space. Indeed we could ask ourselves what speeds we might *expect* interstellar comets to have. The mean speed of the Sun about the galactic centre is about 240 kilometres per second – I write the 'mean speed' since it has a slightly eccentric galactocentric orbit of period about 250 million years, and it also oscillates up and down through the galactic plane with a frequency of one cycle per 60 million years, so that its path is similar to that taken by a carousel horse. Although its orbital speed is 240 kilometres per second, its speed relative to local stars is only about 20 kilometres per second, because they are moving in broadly the same direction. Thus if comets are being ejected from nearby stars, then they would start off with speeds similar to this, relative to the Sun. As they fall into the Solar System these comets would be accelerated by the solar gravitational field, and by the time that they reach the Earth's orbit their speeds would be given by adding the relevant speeds 'in quadrature'; that is, you square each of the speeds, add them together, and take the square root. Those speeds are 42.1 and about 20 kilometres per second, rendering a value of 46.6 kilometres per second. This means that one might expect to see interstellar comets arriving with speeds of four or five kilometres per second above the parabolic limit. And we do not; or at least we have not, so far.

But how often might we expect to observe such comets? David Hughes of the University of Sheffield did some tentative calculations a few years back, based upon an assumption that nearby stars are surrounded by clouds of comets similar to our own Oort cloud, and thus how often comets are pumped into interstellar space, and how often they might be expected to encounter the Solar System subsequently. He found that an observable interstellar comet might arrive about once every four or five centuries, in which case it is not surprising that we have not identified any yet, if they exist.

This means that if one wants to discover an object coming from interstellar space, then one would have to wait for centuries if comets are the only things in which you are interested. Assuming that you are interested in tracking a lot of these, in order to say something about how the Solar System is moving in space, this is not a very attractive proposition. What you need is some type of object with a flux much larger than the ten or so long-period comets that we discover each year.

The answer is to look at meteoroids – even an observer using no equipment except for his/her eyes can see ten or more meteors per *hour* on the average, implying that the overall flux is high. But can we make sufficiently precise measurements of them to determine whether any are arriving from interstellar trajectories? Comets may be observed for months or years in order to accurately determine their orbits, but you only get about a second for a meteor. So there are problems but, as described below, we are just coming to the stage where we can surmount them, and do a proper search for meteoroids coming from interstellar space. But first, because it has been such a long-argued topic, we should look at a little of the history.

One of the great pioneers of studies of small bodies in the Solar System was Ernst Öpik, who died in 1985. Indeed, in 1932 he suggested a structure like the Oort cloud (which was independently invented by Jan Oort in 1950), to which extent maybe we should call it the Öpik–Oort cloud. Whilst Öpik got that aspect of his research correct, there was a related aspect in which he went seriously awry, and that was the question of meteoroids of interstellar origin.

Early indications, for example fireball (bright meteor) observations by the prominent German meteor observer Cuno Hoffmeister and colleagues in the 1920s, had indicated that many meteoroids enter the atmosphere with speeds in excess of the parabolic limit. In fact, since the Earth is moving, the speed observed for any meteor has to be adjusted for our motion. Thus a meteoroid on a parabolic orbit which happens to meet the Earth head-on (which means that it would need an inclination to the ecliptic of 180° and a perihelion distance of one astronomical unit) would be travelling at about 73 kilometres per second as it entered the atmosphere (29.8 plus 42.1 kilometres per second, incremented by being added in quadrature with the Earth's escape speed of 11.2 kilometres per second). Conversely a parabolic meteoroid catching us up from behind (inclination 0°, but perihelion distance as before) would enter the atmosphere at 16.7 kilometres per second. Therefore it is possible that, for a meteoroid arriving from an elliptical orbit, some small error in the determination of the meteor radiant or speed would perhaps lead to a parabolic/hyperbolic orbit being computed, and the mistaken deduction being made that the particle was from interstellar space.

Hoffmeister, with his colleague von Niessl, had estimated that about 80 per cent of their observed fireballs were interstellar in origin. It was against this background that the Harvard College Observatory staged a meteor expedition to Arizona in 1931–33, to check upon Öpik's hypothesis that most meteoroids originated from other stellar systems. The expedition used an ingenious technique devised by Öpik, known as the 'rocking mirror' method. Photographic records were made of the sky as reflected in a mirror which was rapidly rocked in two dimensions, such that any meteor was recorded as a helical trace rather than the usual straight line. From the spacing of the spiral twists the meteor speed could be determined, in principle. Öpik found that 60 per cent of all the observed meteoroids came from interstellar orbits, a result which was later shown to be in error due to a misinterpretation of the data. It took until 1969 before Öpik admitted that a mistake had been made, and retracted the claim that the majority of sporadic meteoroids have an interstellar origin.

Back in the 1930s, however, an interstellar source was still the favoured explanation. At Harvard, Fred Whipple started a series of important meteor observations using twin cameras separated on the ground by many miles, so as to derive better measures of meteoroid orbits. These observations were continued through to the 1950s. A fraction of the orbits were *computed* to be hyperbolic, but the results were consistent with none being from interstellar paths, if timing uncertainties were taken into account.

The tide had turned, therefore, towards all meteoroids being thought of as being members of the Solar System. The supporters of the interstellar origin concept – in particular Öpik and Hoffmeister – fought a valiant rearguard action, but the question was, it seemed, all but settled when radars were first used for meteor observations in the late 1940s. By sending out pulses from a suitable decametre-wavelength radar, it was possible to measure the meteor speed from the way in which the range from the radar site altered for a few very bright meteors, and these did not seem to exceed the geocentric parabolic limit of 73 kilometres per second. A new technique soon became available, though, which allowed speed determinations for large numbers of fainter meteors (see Figure 1). Observations at

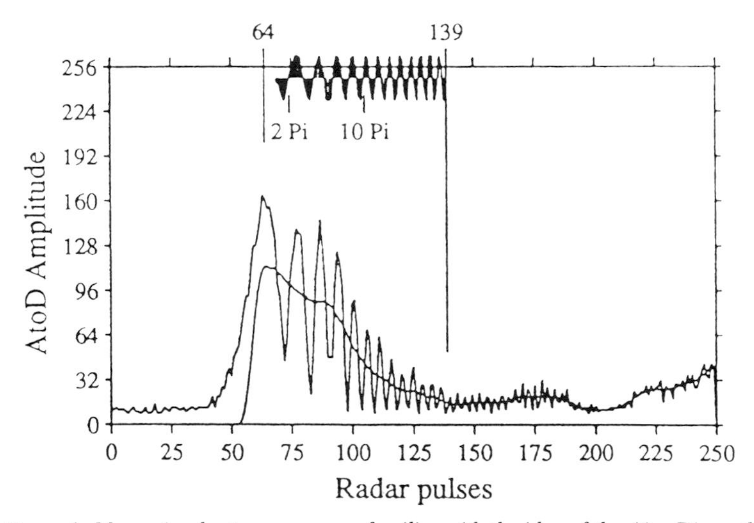

Figure 1. Many visual astronomers are familiar with the idea of the Airy Disc – the concentric set of rings produced by the effect of diffraction in a circular aperture. A straight edge also produces a diffraction pattern, although the alternating bright and dark fringes are then linear, for example, if one pans a microscope over a razor blade illuminated from below. This is called a Fresnel diffraction pattern. Radar meteor echoes are just the same phenomenon, but scaled up – the radar wavelength is much longer than the wavelength of visible light, with the meteor travelling through the radar beam at a range of some hundreds of kilometres acting just like a razor blade moving across the field of view of a microscope. This produces a series of fringes, sampled by the pulsing radar, and the spacing of the fringes (or Fresnel oscillations) allows the meteor speed to be determined using the measured range of the meteor, and knowing the radar wavelength. This meteor was observed using the meteor orbit radar near Christchurch in New Zealand. In this example, the meteor train has mostly decayed away within about a fifth of a second, due to the meteor ionization dispersing into the atmosphere. This is the normal lifetime of a meteor train. The pulse numbers at the bottom are in units of 1/379th of a second, since the radar transmits this number of pulses each second. By taking out the average trend of the echo strength (the solid line through the oscillations) one derives the oscillations only, as shown at the top. Measurement of the oscillation frequency then leads to a determination of the meteor speed.

various installations, in particular at Jodrell Bank in England and Ottawa in Canada, indicated that very few meteor speeds were above that limit: typically 1 per cent of speed determinations appeared to be over 73 kilometres per second, and those could be explained away as being due to measurement errors.

These were only speeds, though. It is possible for interstellar meteoroids to enter the atmosphere at only, say, 20–30 kilometres per second, as foreshadowed above, so that a full answer to the interstellar meteoroid question requires the orbits to be determined. That means that one must measure the radiant as well as the speed.

The first radar meteor equipment capable of measuring meteoroid orbits was that developed at Jodrell Bank in the mid-1950s. Of the 2,000+ orbits measured, about 2 per cent were found to be hyperbolic, but the prevailing wisdom explained these away as being due to random errors. Various other meteoroid orbit surveys over the next two decades – both optical and radar – have led to many thousands of orbits being determined, with up to 25 per cent of orbits being measured as being hyperbolic, but explained away as being erroneous. For example, in the 1960s the Harvard Radar Meteor Project produced 39,000 orbits, about 2.5 per cent of them being hyperbolic. An important point is the bias of the experimenters, which has been against interstellar meteoroids for some decades; for example, the Adelaide radar survey of 1968–69 produced 13.7 per cent hyperbolic orbits, whereas the earlier programme in 1960–61 produced none at all, since the data reduction software was written so as to reject as being 'in error' all apparently-hyperbolic orbits.

All of this chapter so far has been of an historical nature, although one must note that interstellar meteoroids would be of considerable scientific interest, since they would constitute a probe of the interstellar medium, both dynamically and physically, and allow the potential to capture interstellar particles recently arrived in the Solar System. We would have to wait far too long for some interstellar comets to arrive, but we could perhaps collect some interstellar meteoroids if they are passing our way. The question of interstellar meteoroids is, therefore, one which deserves continued study, especially since we now have access to technological capabilities not available when this was such an actively-debated topic a few decades ago. Technological advances always bring the potential to gain significant new insights to problems that might be thought to have been solved, or have merely gone out of fashion.

The first indication that maybe we should reconsider the existence or otherwise of interstellar meteoroids came in the mid-1980s when Jim Jones and colleagues at the University of Western Ontario in Canada developed a sensitive TV system capable of measuring the orbits for very faint meteors; previous optical observations of meteors had all been by using photography, but the newly available solid-state detectors meant that much more sensitive electronic detection systems might be employed. The Canadian team found that about 17 per cent of their orbits were hyperbolic, and, importantly, some of these had speeds well below the 73 kilometres per second limit. Despite this, the prevailing prejudice against interstellar meteoroids meant that little debate was engendered.

The question was re-opened in the past couple of years by information returned by the *Ulysses* satellite. This European Space Agency spacecraft – sent out to pass close by Jupiter to gain a speed boost from its gravitational slingshot effect, then looping up over the pole of the Sun – has on board a dust detector operated by a team under Eberhard Grün of the Max-Planck-Institut in Heidelberg, Germany. That detector has picked up the signature of impacts by tiny dust grains which seem to be arriving from various discrete directions which may be associated with local interstellar sources. For example, one might expect dust to be swept in along the same direction as the flow of interstellar helium, which has been defined by other techniques. The Sun itself is moving in a particular direction relative to the local group of stars, and if these are ejecting dust and meteoroids then a flow from that direction might be expected. The solar motion relative to the nearest O and B stars might also be anticipated to produce a directional source of dust, since these young, hot, fast-evolving stars are thought to throw out large quantities of detritus. The *Ulysses* result is therefore important, although the ability of the dust detector on board to determine the speed and direction of arrival of the dust is limited.

A call to service of ground-based meteor techniques might therefore be in order. Over the past fifteen years I have been involved in the construction and operation of a new, sophisticated, meteoroid orbit radar in New Zealand, along with Jack Baggaley, Bob Bennett and Andrew Taylor. This radar is capable of determining orbits for particles as small as 100 microns, for Solar System orbits; for the higher-speed interstellar particles, the size limit may be as low as 30 or 40 microns. This is close to the sizes of interstellar dust grains studied using other methods, such as infrared satellite-borne telescopes. Previous radars used in orbit determinations have usually measured the meteor speed from the Fresnel oscillations, like those shown in Figure 1. In the past this has meant filming an oscilloscope screen, and later fitting the oscillations using mechanical templates or similar, rather crude, techniques. With our new radar, however, we are able to digitally store the full radar echo profiles and make

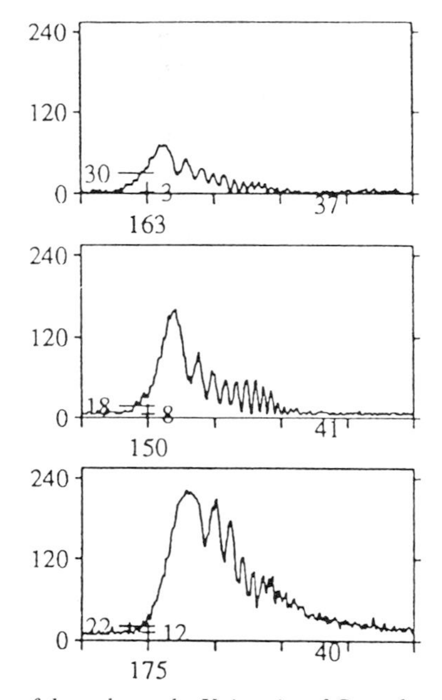

Figure 2. In the case of the radar at the University of Canterbury in New Zealand, we measure the meteor speed from the time-lags between receiver sites separated by some kilometres. Here the times of closest approach to each site are identified as pulse numbers 163, 150, and 175, implying time-lags of up to 25/379 of a second (since there is 1/379th of a second between pulses). In the case of this particular meteor, the Fresnel oscillations on each echo profile also allowed speed determinations to be made, and these were slightly different, reflecting the deceleration of the meteoroid in the atmosphere. Radars used for meteor studies are very different from the dishes seen at airports and so on. The wavelength used in this radar is about 11 metres (frequency 26 MHz), transmission and reception being through wires elevated about half a wavelength above the ground.

rather more precise determinations of the meteor speeds. In fact, the Fresnel oscillation method we use only as a check, since by using three radar sites spaced in a triangular array some kilometres in extent, we are able to measure the meteor speed and radiant from the time lags between the instants at which the meteor makes its closest approaches to each site (see Figure 2). The result is that not only are our orbits more accurate than previous measures, but also we are able to make slight variations in our analyses to see whether orbits initially thought to be hyperbolic might instead be elliptical.

An important indicator to us that hyperbolic meteoroids might be more common than previously believed came when Taylor looked into the problem of *pulse aliasing*. The point is this. Meteor radars measuring speeds from the Fresnel oscillations (Figure 1) are usually pulsed, sending out perhaps 300 or 400 distinct pulses each second. For most meteor speeds this works fine, since the pulse frequency is sufficient to properly sample the oscillations. However, at high meteor speeds it is possible for the oscillations to be undersampled, and then the radar will produce a speed measurement which is wildly wrong; say, a meteor which was actually travelling at 80 kilometres per second being indicated to have a speed of only 25 kilometres per second. By comparing speeds from the time-lag method with those from the Fresnel oscillation technique for particular meteors, Taylor found that this occurs quite often, and is a strong selection effect against very high speed meteors. We therefore reasoned that previous radar results could be in error – an error to which the New Zealand radar is not subject since it can measure the speeds from time-lags.

With this in mind we sorted through the first 165,000 orbits measured – I might note in passing that we now have 350,000 orbits, many times more than in previous meteor orbit surveys – and picked out those which were apparently hyperbolic. This group was then cut down by selecting only those with measured speeds of greater than 100 kilometres per second – far above the geocentric parabolic limit. We then inspected each meteor record by eye in order to exclude all those that might conceivably have been distorted by some atmospheric, or other, effect. This left a core of just 1,508 which we were confident to be extremely high-speed meteors. For each of these we tried adding random experimental errors, to see if they could be elliptical orbits for which our data analysis had mistakenly produced hyperbolic orbits. The answer was that we could not explain them in this way. We also looked at the highest-speed elliptical orbits, and added random errors to those to try to duplicate our high-speed group, again without success. We were therefore convinced that we were detecting real interstellar meteoroids.

But there are more tests that could be tried. If this high-speed group was really due to measurement errors then we might expect them to be randomly distributed through the months of each year of observation. This was far from the case. I have previously mentioned that the Sun orbits the galactic centre at a speed of about 240 kilometres per second. Its direction of motion is pointed towards a point called the *apex*, and this lies at a northerly declination. If you imagine the Earth in its orbit, for six months of each year we will be moving in a direction which has a component towards the apex, and for six months we will be moving away; the fact that the apex is not on the ecliptic means that the Earth's motion is never directly towards it, or away from it. We found that we were detecting more interstellar meteoroids during those months when the Earth is moving in unison with the solar motion, and almost none when moving away from the apex. This is as expected since the incoming speed of interstellar meteoroids (assumedly around 20 kilometres per second when beyond the planetary region) is less than the orbital speed of the Earth (near 30 kilometres per second, as we've seen), so that when the Earth is moving away from the apex, few interstellar meteoroids would be capable of catching the Earth from behind. We also have evidence of discrete sources of interstellar meteoroids, in line with the suggestions made above in discussing the results from the *Ulysses* dust detector.

At the time of writing our data collection in New Zealand is continuing, and soon we will have a valuable data repository of half a million meteoroid orbits. Those have been derived from just five years of observing, whereas rather less than a thousand Earth-crossing comet and asteroid orbits have been determined by mankind over centuries of observing – meteor study is therefore a major contributor to our knowledge of the dynamical structure of the interplanetary, and perhaps interstellar, complex of small bodies. Our analysis of these meteoroid orbits continues, especially with regard to the identification of interstellar particles. When this radar was planned, the question of hyperbolic orbits was not one which we were planning to attack, but it is an important spin-off from a programme intended to investigate many facets of the intriguing interrelationship between comets, asteroids and meteoroids. In fact, because the apex of the solar motion is a long way north in the sky, the radar could not be positioned in many worse places if one wanted to detect interstellar meteoroids. A much better location would be in some temperate northern latitude, perhaps in North America or Europe. We have made many leaps towards proving that interstellar particles are penetrating the Solar System; the next big step will require a new meteor orbit radar to be built somewhere in the Northern Hemisphere. Such a device could answer many important scientific questions, and perhaps show us how to collect our first samples of interstellar material.

Measuring the Universe: The Quest for H_0

SHAUN HUGHES

Now that the Hubble Space Telescope is repaired and giving high resolution images, astronomers are now on the brink of finally obtaining an accurate measure of the expansion rate of the Universe, almost seventy years after it was first observed.

In 1927 Edwin Hubble, an American astronomer working at the 100-inch telescope on Mt Wilson, above the smog of Los Angeles, found from observations of relatively nearby galaxies that the fainter a galaxy appeared to be, the faster it appeared to be moving away from us. Hubble's observations were fundamental in changing our perception of the Universe from a static always-been-there place to a dynamic evolving cosmos, and was the first observational evidence in favour of a Big Bang model for the very creation of the Universe. His result is all the more remarkable when we remember that as late as 1920 the famous debate between the two eminent American astronomers Harlow Shapley and H. D. Curtis over whether the spiral nebulæ were within our own Milky Way galaxy, or as Curtis argued were galaxies in their own right, similar to the Milky Way, but at vast distances. Curtis failed to convince because he didn't have any conclusive evidence. This wasn't provided until 1923, which dates the start of modern cosmology (that is, studying the Universe), when Hubble first detected Cepheids in M.31, and from their brightness he was able to unambiguously establish that M.31 (and by implication all the other spiral nebulæ) were far beyond the bounds of our own galaxy.

And from 1927 to the present, astronomers have been trying to measure just how fast the Universe is expanding. An accurate measure of this rate of expansion, now referred to as Hubble's constant, or H_0 (pronounced H-nought), is vital not only in determining distances to galaxies and hence measuring their total energy, size, and mass, but is required in most cosmological theories to determine the overall density of the Universe, and hence whether it will eventually contract back to a point (the 'Big Implosion') or expand for ever. And perhaps the most fundamental aspect of H_0 is that it is also a measure of the time since the Big Bang, and hence is a measure of the age of the Universe – a vital parameter in determining how long galaxies had to form, and whether theories of stellar evolution give us the correct ages of the oldest stars.

But the irony of the H_0 story is that despite all the effort devoted since 1927 by a veritable small industry of astronomers, so far an accurate value for H_0 has eluded us! Although H_0 is, in principle, a simple concept, only requiring us to know both the expansion velocity of a galaxy and its distance, there are several methods for measuring and interpreting these, and while each method on its own appears to give an accurate value, when compared to the other methods they tend to cluster around two particular values, namely $H_0 = 50$ and $H_0 = 80$ km/s/Mpc. This large difference tells us there must be systematic errors in some, if not all, the various methods. To measure these systematic effects we must have a reliable calibration of all the methods.

Velocities are now relatively easy to measure. All you need is a large telescope and a reasonable spectograph, observe the galaxy for long enough to resolve a particular, easily recognizable atomic line, and compare the wavelength of this line to the wavelength the same line has in a laboratory on Earth (referred to as the rest wavelength). The difference between the observed and rest wavelengths, divided by the rest wavelength, is the redshift of the galaxy. The most distant objects in the Universe appear to be quasars, which are most likely very young galaxies with extremely bright cores, and which have redshifts typically between 1 and 4. The nearest galaxies only have redshifts around 0.001, and so their redshifts are most often quoted as actual velocities (a redshift of 0.001 is equivalent to a velocity of 300 km/s). Measuring redshifts is particularly easy for spiral and irregular galaxies, because they still have lots of hydrogen gas and recent star formation. These very hot, bright young stars heat the gas around them, and the hydrogen atoms in the gas cool by emitting characteristic radiation in very narrow but bright emission lines, the one most easy to measure being that at a rest wavelength of 656 nm, called H-α (H-alpha). For elliptical galaxies, where most of the gas is either lost or completely ionized, velocities are measured from absorption lines, mainly those of magnesium and calcium. Although velocities are easy to measure, it is their interpretation that is a problem. This is because

Figure 1. M.100. These images of the nucleus and surrounding area of the giant spiral galaxy M.100, in the Virgo cluster of galaxies, shows the remarkable resolution achieved by the repaired Hubble Space Telescope (above) compared to a ground-based image obtained from a large telescope under good seeing conditions (below) (Photograph courtesy of NASA/JPL).

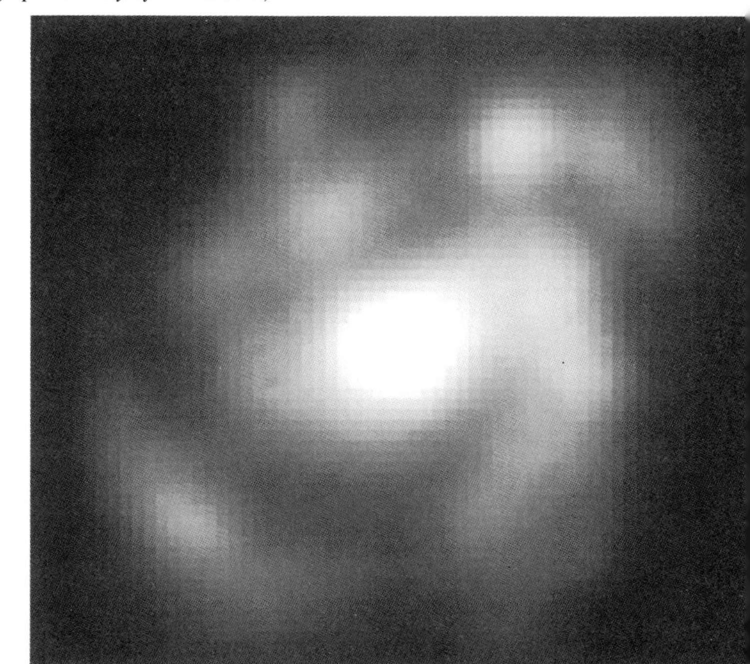

the motions of galaxies are not only caused by their cosmological expansion (also called the Hubble flow), but also by gravitational interactions with neighbouring galaxies (giving rise to so-called peculiar velocities). Because these peculiar velocities are mostly less than about 200 km/s, this means that to measure what is mostly a real Hubble flow, we need to measure distances to galaxies with total velocities (Hubble plus peculiar) greater than about 1000 km/s, which are found at distances beyond about 15 Mpc. (As most astronomical distances are ultimately based on the parallax distance to the nearest stars, distances are hence measured in terms of parsecs (pc), where 1 pc is the distance a star would have with

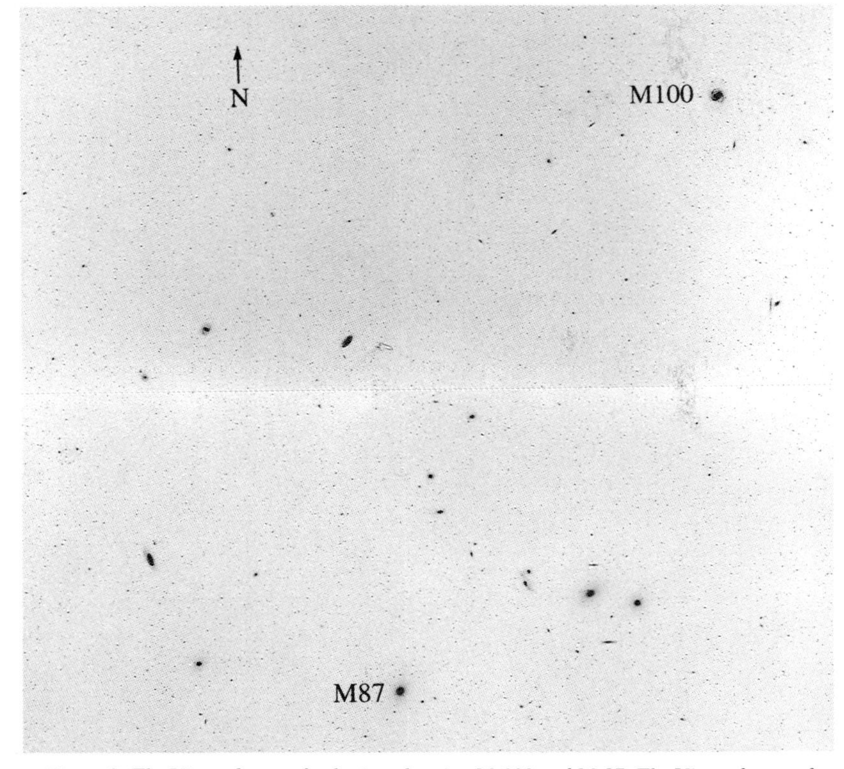

Figure 2. The Virgo cluster of galaxies, showing M.100 and M.87. The Virgo cluster of galaxies contains over 2500 galaxies, centred about the giant elliptical M.87. M.100 is about 4° away to the northwest (A section of one of the plates from the First Palomar Oschin Schmidt Survey).

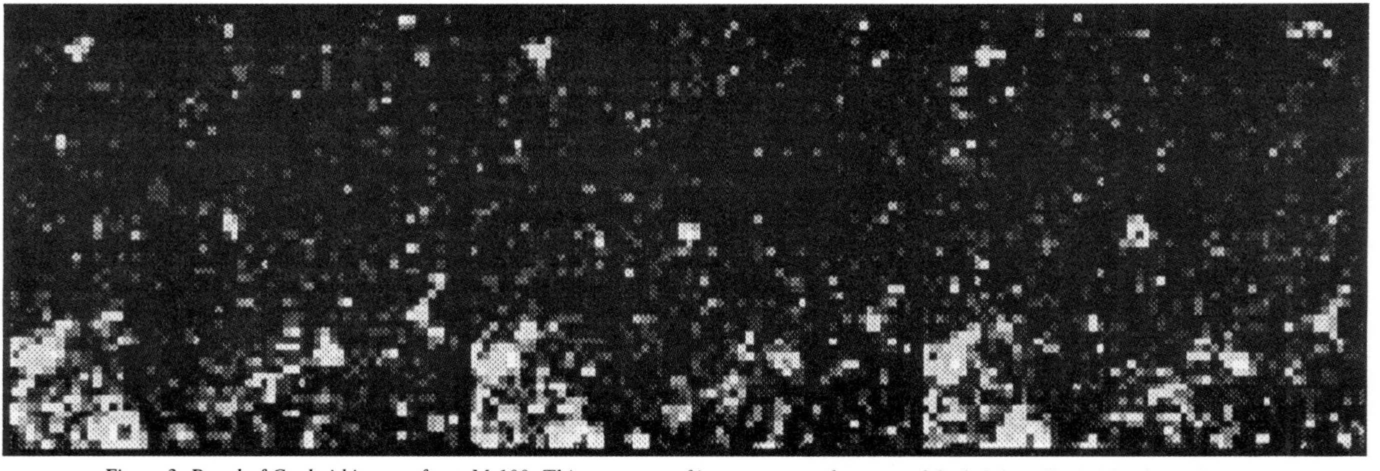

Figure 3. Panel of Cepheid images from M.100. This sequence of images, centred on one of the brighter Cepheids, shows how its brightness is varying, from May 9 (left) to May 4 (middle) to May 31 (right) (credit: L. Ferrarese, Johns Hopkins University, Baltimore, USA).

1 arcsecond of parallax, which is equivalent to 3.26 light-years, or 31 million million km. Hence 15 Mpc is about 50 million light-years, or 465 million million million km!)

While velocities are easy to measure, distances are certainly not. In fact, measuring distances to anything in astronomy is extremely difficult, let alone the distances to far away galaxies. So how do we measure distances? Well, for the very nearest 10,000 stars (these are all within 30 parsecs) we can use the only direct method, called parallax, which employs simple trigonometry, using the Earth–Sun distance as a baseline of a very long isoceles triangle formed with a star at the apex, then measuring the 'height' of the triangle via the two angles on the Earth–Sun base, which are measurable for stars within 30 parsecs. Distances to stars beyond this generally rely on knowing the intrinsic brightness of a particular type of star. Mostly these particular stars are periodic variables – stars whose brightness varies over a constant period – and the most reliable of these are the Cepheid variables. Cepheids are yellow supergiants with periods of between about 3 and 80 days, with the longer periods belonging to the more massive and hence the more luminous Cepheids (a 40-day Cepheid is about 10 times the mass of the Sun, and about 6000 times more luminous). This naturally gives rise to a period-luminosity (P-L) relation, in which the luminosity increases linearly with the logarithm of the period. Once the P-L relation is reliably calibrated, then measuring the period of any Cepheid will immediately tell you its intrinsic brightness, and so the difference between this and its observed (or apparent) brightness will give you the distance to the Cepheid and hence the distance to the galaxy in which the Cepheid belongs. Ground-based telescopes have been used to detect Cepheids in the Magellanic Clouds, the Andromeda spiral galaxy (M.31) and as far away as M.81 and M.101, and most recently, using techniques to compensate for fluctuations due to seeing, to NGC4571 in Virgo. However, because the Earth's atmosphere blurs stellar images, and the further away a galaxy is the more crowded its stars appear to be, it is extremely difficult to detect more than a few Cepheids in the more distant of these galaxies, and there is always a problem of completeness which may bias the resultant distance (the brightest Cepheids are naturally the most likely to be seen, so unless you have a large number of Cepheids to compare to a known P-L relation, you will end up measuring a distance which is shorter than the true distance).

For these reasons a large fraction of the Hubble Space Telescope

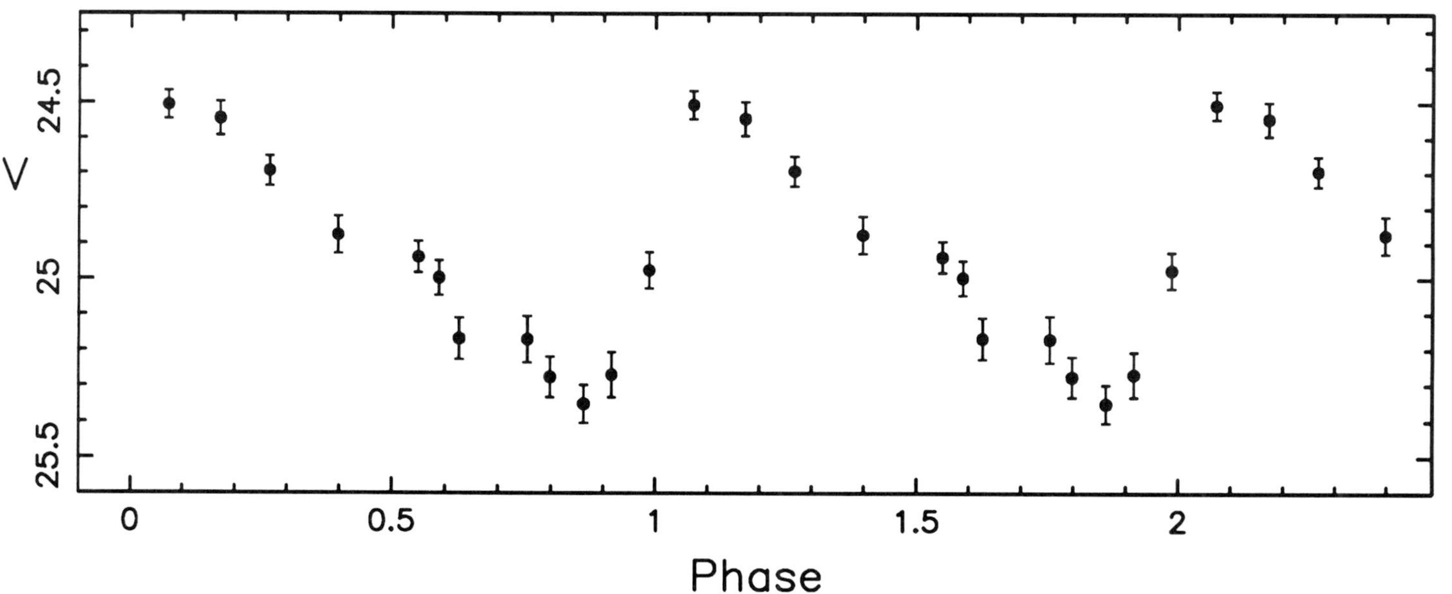

Figure 4. Cepheid light curve. Sophisticated computer programs allow us to measure both the brightness of each star and match all stars from one epoch to another, thereby allowing us to build up the light curves for every star in the field. We then select those which are varying periodically, and from these search for the brighter Cepheids, which tend to have the above characteristic shape and periods between 10 and 80 days. This Cepheid has doubled in brightness over a period of 51.3 days (credit: W. Freedman, Carnegie Observatories, Pasadena, USA).

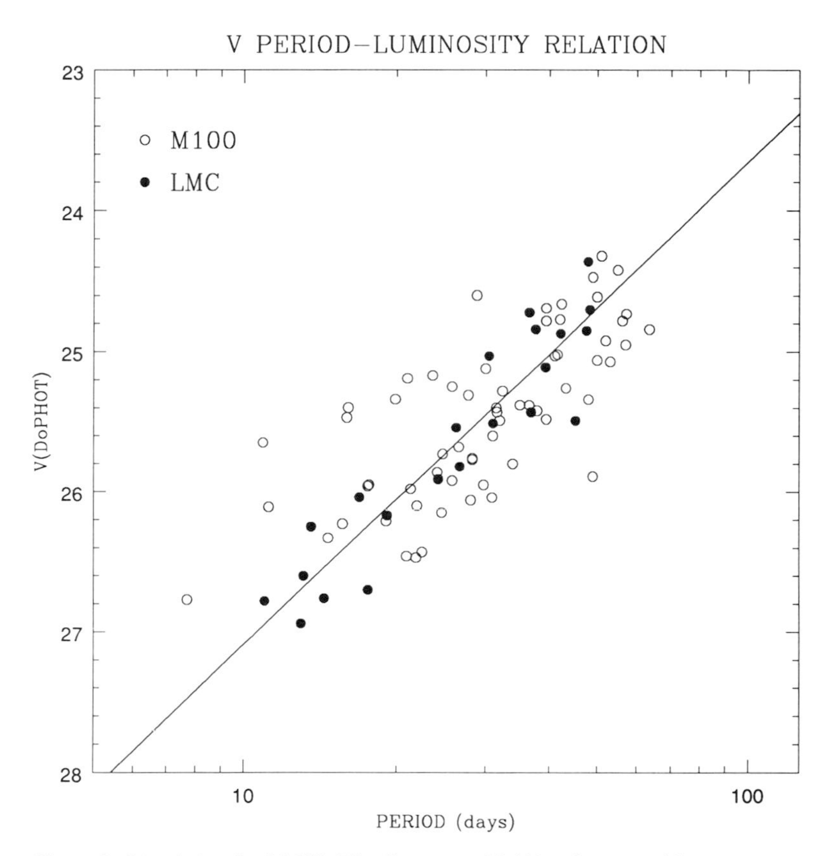

Figure 5. PL relation for M.100. The distance to M.100 is determined by comparing the straight line relation between luminosity (brightness) and the logarithm of the period for Cepheids in M.100 with a standard set of Cepheids in the Large Magellanic Cloud, whose distance is relatively well known. The open circles represent the M.100 Cepheids, and the filled circles and straight line are the LMC Cepheids and the LMC period-luminosity relation, shifted in magnitude to agree with the M.100 sample. The amount of shift gives us the relative distance between the LMC and M.100 (credit: L. Ferrarese, Johns Hopkins University, Baltimore, USA).

(HST) observing time has been assigned to the quest for an accurate measurement of H_0, and has been given the status of a Key Project. (Much of the original motivation for the Extragalactic Distance Scale Key Project was provided by Marc Aaronson, who tragically died while observing at Kitt Peak in 1987.) HST is ideal for detecting

Cepheids, not just because of the obvious advantage that it is above the atmosphere and has a seeing-free resolution of 0.1 arcseconds, but it is also above the clouds and orbits the Earth every ninety minutes, which means that it is possible to reliably schedule observations which will optimally sample the Cepheids' light curves, and therefore gives accurate periods.

We now have good reason to believe that we are on the threshold of finally nailing down H_0. This is because in 1994 we were successful in measuring the distance to M.100, a galaxy in the Virgo cluster of galaxies, at a distance of 17.1 Mpc. This was achieved by Wendy Freedman and the rest of the Key Project team (see *Nature* 1994 Oct. 27) using the repaired HST. Although this is almost the limit to which even HST is able to detect Cepheids, distances to galaxies beyond Virgo can be measured using a variety of ingenious methods developed over the last twenty years. Some of these secondary distance indicators use the relation between the internal motions of

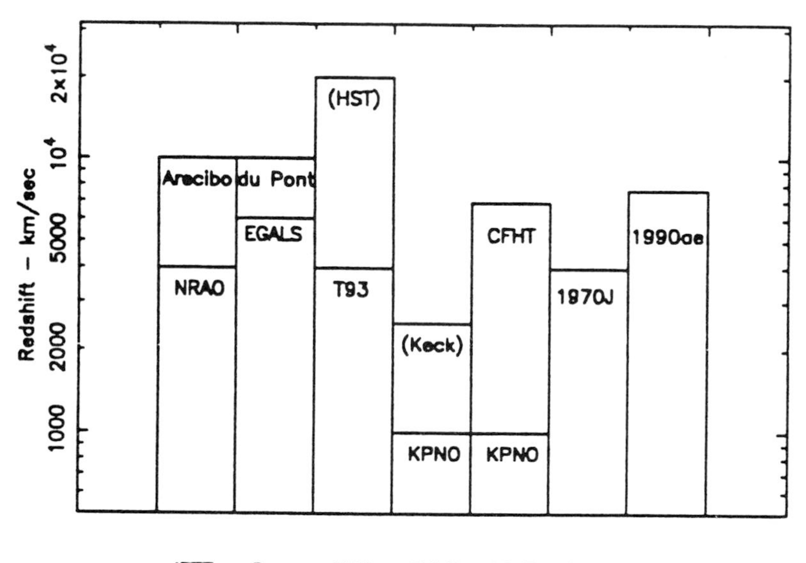

Figure 6. Redshift distances reachable by secondary distance indicators. The redshifts that are reachable, using various telescopes, by the secondary distance indicators we are hoping to calibrate. These velocity redshifts may be converted to a distance by dividing by your favourite value of H_0 (currently somewhere in the range 50 to 100 km/s/Mpc).

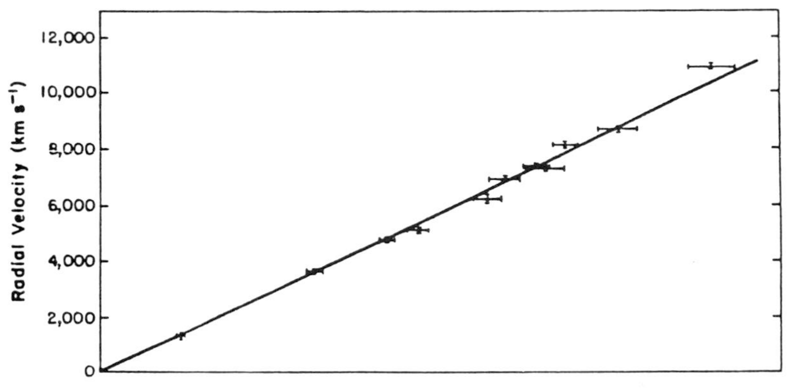

Figure 7. Uncalibrated Tully–Fisher relation. The aim of the Extragalactic Distance Scale Key Project is to provide reliable distances to enough galaxies that we may calibrate various secondary distance relations, such as the one shown here for the IR Tully–Fisher relation, for 11 galaxy clusters. This calibration will then provide the horizontal distance scale. But even in its uncalibrated form, the dispersion about the straight line relation is remarkably small. The lowest redshift cluster is the Virgo cluster, corrected for peculiar velocity.

stars within a galaxy and its total mass, and hence total light, and include the Infrared Tully Fisher relation (IRTF), which measures distances via a relation between the rotational velocity of spirals and their integrated light, and the $D_n - \sigma$ relation for elliptical galaxies, which compares the velocity dispersion of the stars within a galaxy with the angular size of the galaxy. Other methods measure the light from either a particular population of stars whose intrinsic brightness is known, such as the surface brightness fluctuations method (SBF) and the planetary nebulæ and globular cluster luminosity functions (PNLF and GCLF), or of a single very luminous event – the supernovæ explosions of white dwarfs (type Ia, denoted as SN Ia) and supergiants (type II = SNII–EPM). (There are also other primary methods which attempt to measure H_0 directly from perturbations to the cosmic microwave background and from distant gravitational lenses, but these are still highly uncertain and model dependent.)

The goal of the Key Project is thus to measure reliable (Cepheid) distances to about twenty 'nearby' galaxies, and use these distances to calibrate the secondary distance indicators. These calibrated secondary distance indicators will then allow us to measure accurate distances out to nearly 300 Mpc, and thus to measure an accurate value of H_0. We already know that most of these secondary distance indicators give reliable relative distances, and we already have Cepheid distances to 3 of our 20 galaxies. All we need to do now is to measure the Cepheid distances to the remaining 17 galaxies, over the next two years, and provide a calibration for each of them, and hence put a real distance scale on the Universe.

The Flammarion Observatory, Juvisy: past, present, and future

RICHARD McKIM

'N'est-il pas étrange que les habitants de notre planète aient presque tous vécu jusqu'ici sans savoir où ils sont et sans doubter les merveilles de l'Univers?' *

CAMILLE FLAMMARION (1842–1925)

The past

The astronomical history of Juvisy is mostly the life story of Camille Flammarion, France's most famous popularizer of that science. He was much more than just the French contemporary of England's Sir Robert Ball, also a well-known lecturer and the author of many popular books. This, then, is the story of the observatory that Flammarion established at the town of Juvisy.

Flammarion was born on February 26, 1842 and he could trace his interest in astronomy back to an observation of an annular solar eclipse when he was just five years old. As early as 1858 we find him working at the Paris observatory, as a junior computor at 50 francs a month, but the work did not suit him. It was the poetic aspect of astronomy that appealed to Flammarion, and as an antidote to the 'autocratic rule and unconciliatory manner' of Le Verrier (in the words of one obituarist), he wrote an imaginative story of a journey to the Moon. This was never published, but he soon enjoyed enormous success with a popular book *La Pluralité des mondes habités* (1862). Such writing was not to his superior's taste, and Flammarion was dismissed, but he found another post at the Bureau des Longitudes. Flammarion was also writing for the most important Paris daily paper, *Le Siècle*, and soon began lecturing at the École Turgot. Other books quickly followed. Le Verrier was to recall Flammarion in 1873, and place at his disposal the great

* (translation) 'Isn't it strange that the inhabitants of our planet have almost all lived until now without knowing where they are and without suspecting the marvels of the Universe?'

equatorial of the observatory's east tower for the latter's investigations of binary stars. From then until 1879 Flammarion was cataloguing these stellar systems, and examining some 11,000 previously observed objects to look for evidence of orbital movements.

While still a young man, Flammarion was becoming well known as a writer and lecturer, and founded his *Annuaire Astronomique*, an almanac, in 1864. One of his best-sellers appeared in 1877, published by his brother Ernest. Entitled *Astronomie Populaire* (*Popular Astronomy*), it went through many editions (over 130,000 copies were printed) and made a large amount of money for both brothers: by 1893 Flammarion thought it had earned him over $25,000 in royalties. By this time Flammarion had many admirers, among them a Monsieur Méret, of Bordeaux. In 1882 Méret offered Flammarion his estate at Juvisy-sur-Orge, not far from Paris, and Flammarion accepted. The estate consisted of a large house or château of three floors, complete with stables and servants' quarters, set in a large park. It was an ancient residence known as the 'Cour de France', and a place where the kings of France had customarily rested on their annual journeys from Paris to Fontainebleau, from Charles VI till Louis Philippe. It was in a room of the same château that the Emperor Napoleon Bonaparte learnt, on March 30, 1814, the news of the capitulation of Paris and the downfall of his Empire.

Flammarion soon placed a fine 9-inch (24-cm) Bardou refractor under a dome on the roof, and this telescope and others were extensively used over the next half century. Flammarion's writing career continued apace, and in 1887 he founded the French Astronomical Society (*Société Astronomique de France*, SAF). He took a fatherly interest in the SAF, as its first president and editor of its *Bulletin, 'l'Astronomie'*. The SAF *Bulletin* became the principal medium through which the results of his observatory were published. The planet Mars was a special source of fascination to him, and he published two great volumes describing in detail all that had then been written on the subject. They are rare books today, and I am glad to have my own copies through the kindness of Professor Jean Dragesco. Such were Flammarion's special interests that his planetary papers were filed under 'Mars', 'Jupiter', and 'Planets other than Mars and Jupiter'. Famous astronomers, such as E. E. Barnard and Percival Lowell, visited Camille Flammarion at Juvisy. Flammarion wrote widely, sometimes for a strictly scientific

Figure 1. Camille Flammarion with his 9-inch refractor at Juvisy. (Juvisy Observatory archives) (Figures 1 and 2 reproduced by courtesy of M. Pernet. No date, probably from the 1890s.)

Figure 2. The exterior of the Juvisy Observatory, from an old postcard.

readership, but most of his books (there were more than sixty eventually) were aimed at the layman, such as his famous romantic novel, *Uranie*. His interests were not confined to astronomy, and he made many scientific balloon ascents. Flammarion was also intensely interested in the occult, and held many séances in his apartment. At Juvisy he also carried out experiments to see how plant growth was affected by screens of coloured glass. In 1894 an agricultural climatological station was annexed to the observatory.

Although a good observer with keen sight, Flammarion seems to have preferred to pay one or more assistants to carry out the programme of the observatory. His interests were in the mapping of planetary surfaces, measuring and cataloguing double stars, and in observations of the various clusters and nebulæ. An early associate was the Greek-born planetary observer Eugène Michael Antoniadi, who later became the greatest authority on the planets Mercury and Mars. Antoniadi worked at Juvisy between 1893 and 1902.

It was to be Ferdinand Quénisset who gave a lifetime of service to Juvisy, and he was still working there in the late 1940s. Quénisset was also interested in planetary astronomy, but he is perhaps best remembered for his astronomical photographs. In addition to such feats as photographing the surface details on Mercury, Quénisset took superb photographs of comets with a triplet Zeiss lens of 6½-inch (17-cm) aperture, attached to the mounting of the 9-inch. He was also the discoverer of two comets.

Flammarion was twice married, first to a widow, *née* Sylvie Petiaux-Hugo in 1874. In May 1893 Robert H. Sherard, a newspaper reporter, visited the Flammarions at their fifth-floor apartment in the Rue Cassini in Paris. The interview which they gave was published in the *San Francisco Call* (I am grateful to Dr William Sheehan for sending me a copy of this interview), and it is a delight to read how Sylvie described her husband's routine. She said:

> 'Flam is an extremely methodical man. He gets up regularly every morning at seven o'clock and spends quite a long time over his toilet. Savants as a rule are a very untidy set, and Flam is an exception to the rule. . . . At a quarter to eight every morning he has his breakfast, with which he always takes two eggs. From eight to twelve he works. At noon he has his déjeuner, over which he spends a long time. He is a very slow eater. From one to two he

Figure 3. E. M. Antoniadi, en route to Norway in 1896 to observe a total solar eclipse on behalf of the Flammarion Observatory. (British Astronomical Association archives.)

receives, and as he is constantly being consulted on all sorts of questions by Parisian reporters, he is usually kept very busy during this hour. From two to three he dictates letters to me. . . . At three o'clock he goes out and attends to his business as editor

Fig. 65.

[4mm = 1".]

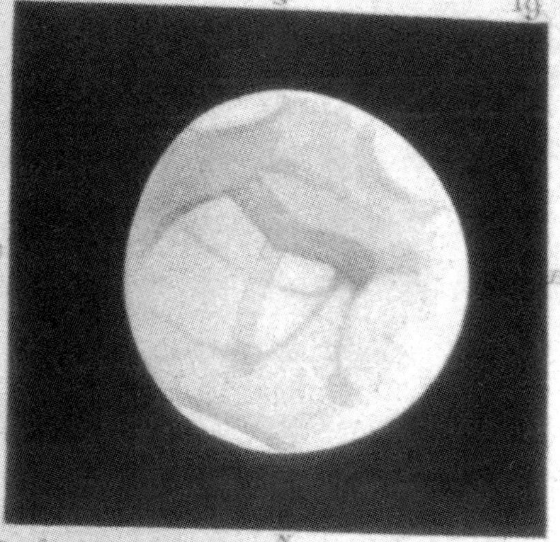

S 19

O E

N

MARS, le 23 octobre 1896, à 14h 0m. λ = 342°. φ = + 2°.7.

Fig. 66.

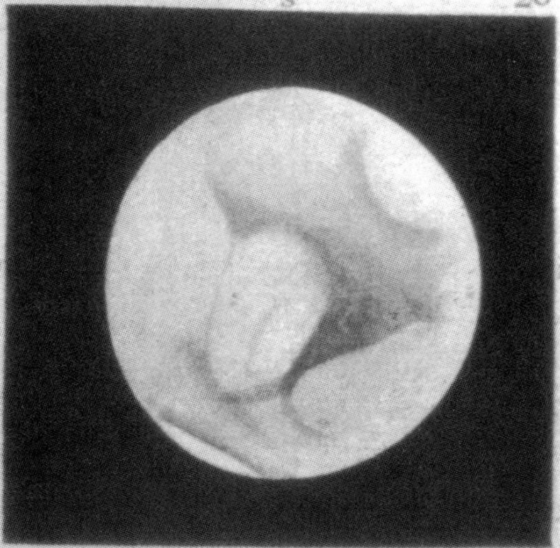

S 20

O E

N

MARS, le 29 octobre 1896, à 12h 45m. λ = 269°. φ = + 2°.7.

E. M. Antoniadi

Figure 5. An early, undated photograph of F. Quénisset (1872–1952). (Juvisy Observatory archives, courtesy M. Pernet.)

of the monthly magazine which he founded and to his duties as a member of various societies. He is back home again at seven-thirty, when he has dinner and spends the rest of the day in reading. He is a great reader, and tries to keep himself *au courant* with all that is said on the important topics of the day. At ten o'clock he goes to bed, for he is a great sleeper.'

Figure 4. Mars drawings by Antoniadi with the 9-inch refractor, 1896. (Juvisy Observatory archives, courtesy M. Pernet.)

'But when,' I asked, 'does Mr. Flammarion observe the stars?'

'Oh, that is his winter programme,' said his wife, 'that I have been describing. It is in the summer when he is down at Juvisy that he continues his studies in astronomy, that is to say, from May to November . . .'

We also learn that Flammarion never smoked: 'It is impossible to observe the stars with a cigarette in one's mouth,' he told Sherard.

The First World War was a terrible blow for Camille and Sylvie, who were both pacifists. Sylvie died in 1919, but Camille married again. His second wife, Gabrielle Renaudot, was an astronomer in her own right, and she had already done observational work at Juvisy. After Flammarion's death on June 3, 1925 Gabrielle carried on his work with the SAF, acting as general secretary and editor of the *Bulletin*. She lived on her own at Juvisy, with her elderly maid for company. She was, Professor Dollfus told me, a courageous and extremely knowledgeable woman, and helped to keep the SAF going through the difficult years of the Second World War. She was still general secretary at the time of her death in 1962.

The present

When Gabrielle Flammarion died, leaving no heir, she bequeathed the estate to the French Astronomical Society. As we shall see, this extremely generous gift (the wish of Camille himself), has not been without some difficulties for the SAF.

In October 1992 I visited Juvisy for myself. I have always been fascinated by the career of E. M. Antoniadi, Flammarion's one-time assistant. I wanted to see the observatory where he first worked, and some of his original drawings. This visit was possible owing to the kindness of Professor Audouin Dollfus of the Meudon Observatory, and the curator of Juvisy, Monsieur Jacques Pernet. It was a beautiful autumn day; the tranquil park formed a stark contrast to the busy main road which runs right past the observatory gate. From outside, the buildings looked just as they do on old postcards of Juvisy. The gardens, now open to the public, are well-kept, and we were able to pay homage to Flammarion, who is buried there with his two wives. It was for me a poignant moment.

Figure 6. Comet Morehouse 1908 photographed by Quénisset. (From a booklet published in memory of Flammarion: Aux Amis de Camille Flammarion, *1926.)*

Figure 7. Camille Flammarion, in old age, in his astronomical library at Juvisy. (From the Bulletin *of the* Société Astronomique de France, *December 1975.)*

The tomb is covered by a flower-bed containing a low hedge in the shape of a five-pointed star. Behind the tomb there is a granite column with a plaque. The SAF hold an annual ceremony here, to celebrate the astronomer's life and work.

Walking back to the house we saw the place on the wall where there had once been a sundial, and ascended by a little turreted tower to the roof. Flammarion had this tower built in 1894 so that his assistants, on arriving for night duty, could ascend to the observatory without having to pass through his private apartments. On entering the observatory last October, however, it was immediately apparent that a lot of effort will be necessary to put the equipment back into working order. Why? Let me explain. In 1944, Allied aircraft were bombing the French railways prior to D-Day. Juvisy, being an important railway centre, was a prime target. So heavy was the bombardment that, although the observatory received no direct hit, the foundations of the ancient property were shaken. Afterwards, it became necessary to shore up some of the ceilings with scaffolding, but Gabrielle continued to live under such conditions until her death. The SAF leased the estate to the municipality of Juvisy for a period of 99 years, the arrangement being that the SAF could continue to use the observatory. In the 1980s the authorities began tests upon the walls and ceilings to assess the state of the building so that renovations could be carried out, the idea being to create in the château a museum of Flammarion and of Juvisy. The furniture of the house went into store, while M. Pernet (supported by Dollfus and other officials of the SAF) did a magnificent job in removing Flammarion's library of over 10,000 volumes to a secure annexe.

I had the pleasure of a guided tour around the nearly empty house, with Messrs Dollfus and Pernet as my expert guides. In spite of the general state of untidiness due to the effects of the structural survey, one felt the presence of a strong sense of history. The 9-inch is still on its mount on the roof, but it is not safe to use. Portholes of coloured glass admitted the light. In a separate room was Quénisset's darkroom, with just an old sink in the corner. We stood in Flammarion's library, a melancholy sight, with the empty, carved bookcases and most of the original wallpaper, the autumn sunlight streaming in through the beautiful coloured-glass windows. What personalities and conversations those walls must have seen and heard! We saw the reception rooms, bedrooms, and bathroom, and a room where Flammarion once displayed a great collection of scientific instruments and curios, presented to him by the many visitors he received.

We examined the fabulous astronomical library in the annexe, containing many ancient rare books as well as a very complete collection from Flammarion's own lifetime. It is a superb resource for the astronomical historian, and contains Flammarion's private papers and correspondence. The archives also contain four manuscript notebooks by Antoniadi, and we carried them outside so that I might make colour slides of some of the pages in daylight. The notebooks are filled with a rich collection of watercolours, pastels, and pencil drawings, mostly of planets and nebulæ, and I was very glad to have had the chance to examine Antoniadi's work at first-hand.

The future

What of the future of Juvisy? Sadly, although funds were available for the survey of the building, they have now run out. It is hoped that the château will eventually be converted into a museum, and that the original furniture of the house can be replaced. Once that is done, members of the SAF will be able to renovate the upper floor and put back into working order the 9-inch refractor. Such things probably lie well in the future. Personally I hope that some sort of appeal fund will be launched in France. Let us hope that one day we may see Juvisy restored as the great centre of astronomy that it once was.

Looking back over a hundred years of progress in astronomy, one cannot fail to be impressed by the sheer magnitude of Flammarion's achievement. The founder of a national astronomical society and the first French astronomical journal, a great writer and philosopher, Flammarion reached a massive popular audience, while his scientific works made him a leading figure of his time. As Professor Dragesco wrote to me recently, generations of French astronomers have been brought up on Flammarion's books. They, too, owe him a great deal.

Bibliography

A contemporary Obituary Notice of Flammarion can be found in: *Monthly Notices of the Royal Astronomical Society*, volume 86, pages 178–180 (1926).

Further details about Flammarion and Juvisy can be found in: *Bulletin of the Société Astronomique de France*, volume 89, 1975 December (special Flammarion issue); and in the article by J.-C. Pecker and J. Pernet in the same periodical, volume 101, pages 331–342 (1987).

An excellent biography of Camille Flammarion has just been published: P. de La Cotardière and P. Fuentes, *Camille Flammarion*, editions Flammarion, 1994.

Many of Quénisset's photographs appear in: *The Flammarion Book of Astronomy*, Allen and Unwin, 1964.

Antoniadi's connection with Juvisy is reviewed in: McKim, Richard, 'The Life and Times of E. M. Antoniadi (1870–1944)', *Journal of the British Astronomical Association* volume 103, pp. 164–170 (1993 August) and pp. 219–227 (1993 October).

The Moon Hoax

or

Great Astronomical Discoveries lately made by Sir John Herschel at the Cape of Good Hope

J. E. KENNEDY

The Moon, the only natural satellite of the Earth, is the nearest celestial object to our planet. Studies have been made frequently to determine any possible effects the Moon might exert on human behaviour. Evidence in support of such effects is neither satisfactory nor convincing. Myths or legends connected with the Moon are no longer in vogue. To an observer on the Earth, the Moon is of the same apparent size as the Sun. Under appropriate circumstances the Moon provides us with an excellent occulting disk, leading to a total eclipse of the Sun, 'one of the most impressive sights that nature offers to the eye of man.'

The interest of the public is easily stimulated by directing attention to events in the sky. In 1833 a British astronomer, Sir John Herschel, travelled to South Africa where he erected a telescope to carry out a systematic survey of the southern skies. Observations of the Moon were not included in his plans. Within two years, and well removed from his observing site, a newspaper in New York revealed to the public the 'Great Astronomical Discoveries' Sir John Herschel had made at the Cape of Good Hope.

The telescope Sir John erected in South Africa, a standard reflector of the period, was similar to one used in England by his father, Sir William Herschel. The accounts published in the New York paper gave the younger Herschel credit for many discoveries, some well beyond the capabilities of his telescope and impossible to carry out in the interval of a year and a half. These accounts captivated the interest of the public in the city of New York who did not question the validity of the claims.

The article in *The Sun* of New York on August 25, 1835 indicated that the younger Herschel had:

— made the most extraordinary discoveries in every planet of our solar system,
— discovered planets in other solar systems,
— obtained a distinct view of objects in the Moon,
— affirmatively settled the question whether this satellite be inhabited, and by what orders of beings,
— firmly established a new theory of cometary phenomena,
— solved or corrected nearly every leading problem of mathematical astronomy.

A distinct view of objects on the Moon, and settling the question of whether the satellite was inhabited, could be achieved only, as the account stated, with 'a telescope of vast dimensions and an entirely new principle.' The reflecting telescope transported to South Africa had a clear aperture of 18¼ inches and 20 feet focus. The great reflector used in England by the elder Herschel was equipped with a mirror 4 feet in diameter and 40 feet focal length, capable of a magnifying power of 6000 times. Taking the mean distance of the Moon from Earth as 240,000 miles, at this magnification the observer could see details on the Moon as though it were a mere 40 miles distant. New approaches and extensive modifications would be required if the lunar surface were to be scrutinized and minute details revealed.

In collaboration with Sir David Brewster, John Herschel was reported to have formulated and tested the radical changes made to the telescope at least a couple of years before leaving for South Africa. The greatly enlarged image of the Moon would be projected on a viewing screen. Additional light would be fed into the normally faint image using a principle borrowed from a standard procedure employed in microscopes. A reversion would be made to an earlier design of refracting telescopes, that is, the tube of the instrument would no longer be needed.

As this fantastic account of 'Great Astronomical Discoveries' was copied and reprinted in other countries, writers did not hesitate to modify telescope design still further. *The History of the Moon*, published in London, started by converting the reflector to a refractor consisting of 'an Object Glass weighing Seven Tons, with a magnifying power of 42,000 times.' A magnificent objective by the most modern standards; the Moon would appear still nearly six miles distant from the observer.

The series of articles covering 'Great Astronomical Discoveries' was published in *The Sun* during the last week of August 1835. Authorship of these accounts has been attributed to Richard Adams Locke (1800–1871). Emigrating from his birthplace of England in 1832, Locke joined the staff of the newspaper a few months prior to the appearance of these reports. He was capable of writing very plausible stories, thinly sprinkled with attesting facts, which he cleverly filled with astronomical terms and the names of scientists. Individuals of his own creation provided a support group for Herschel at the Cape Observatory.

Readers were completely deceived, unaware that a tremendous hoax was being perpetrated on them. Sales of the newspaper soared to outstanding numbers. Locke described the inhabitants of the Moon so vividly and in such detail that illustrators had a simple task to present an overall view of the lunar landscape, close-ups of Biped Beavers, Lunar Quadrupeds, and Human-Like Inhabitants. All of these features and creatures were reported to have been seen through the telescope of Sir John Herschel:

> BIPED BEAVERS:
> resembles the beaver of the earth in every other respect than its destitution of a tail, and its invariable habit of walking upon only two feet. It carries its young in its arms like a human being and moves with an easy gliding motion. Its huts are constructed better and higher than those of many tribes of human savages, and from the appearance of smoke in nearly all of them, there is no doubt of its being acquainted with the use of fire . . . never seen except on the borders of lakes and rivers.
>
> LUNAR QUADRUPEDS
> we beheld continuous herds of brown quadrupeds, having all the external characteristics of the bison, but more diminutive. . . . It had, however, one widely distinctive feature . . . namely a remarkable fleshy appendage over the eyes, crossing the whole breadth of the forehead and united to the ears . . . lifted and lowered by means of the ears. . . . This was a providential contrivance to protect the eyes . . . from the great extremes of light and darkness to which all inhabitants of our side of the moon are periodically subject.
>
> HUMAN-LIKE INHABITANTS OF THE MOON
> we were thrilled with astonishment to perceive four successive flocks of large winged creatures . . . descend with a slow even motion from the cliffs . . . and alight upon the plain. . . . They were like human beings, for their wings had now disappeared, and their attitude in walking was both erect and dignified. . . . They averaged four feet in height, were covered, except on the face, with short and glossy copper-colored hair, and had wings composed of a thin membrane, without hair, lying snugly upon their backs from the top of the shoulders to the calves of the legs.
> (designated as Vespertilio-homo, or man-bat)

Locke may have been an observant naturalist, familiar with the features of the birds and animals he described. He attested the lunar 'sheep would not have disgraced the farms of Leicestershire.' To an extent this and other references revealed his origin and background, a matter he realized it was as well to conceal. For an accurate picture of the water birds seen on the Moon, Locke had only to turn to a drawing of a constellation in a book like Blunt's *Beauties of the Heavens*. The sources of the astronomical information used by Locke have not been identified with certainty.

The separate articles written by Locke in *The Sun* were collected and published as a pamphlet. Sales of some 60,000 copies of this were reported. The only known copy of this extensive press run is housed in the Library of Congress, Washington, DC. Reprints with additions and revisions, and translations of the pamphlet soon appeared in the UK and several European countries. With an explicit descriptive text available, illustrators and engravers took full advantage of the possibilities open to them. An Italian publication included an engraving entitled the 'Flying Diligence', an airship in which the lifting and motive power were provided by Vespertilios.

Sir John Herschel's *Results of Astronomical Observations Made During the Years 1834, 5, 6, 7, 8, at the Cape of Good Hope* was published in London in 1847. A direct comparison of the headings of the seven chapters in this monumental treatise with the six claims presented by Locke indicates only two of the claims relate to work performed at the Cape. Studies were carried out by Herschel on the satellites of the planet Saturn and the 1835 return of Halley's Comet was observed. Thus credit can be given to Locke for having one-third of his claims substantiated.

What was Herschel's reaction to the Moon Hoax? At the time he was very much in the spotlight as an astronomer. His friends and colleagues did not appear hesitant in making him aware of what had taken place in the world of journalism with the publication of 'Great Astronomical Discoveries'. In May 1836 Robert Treat Paine of Boston wrote to Herschel:

> 'In September last, there appeared in a newspaper, the Sun, published in New York, but not held by the inhabitants of that City in high repute, an account of some discoveries in the Moon, alleged, to have been recently made by yourself, which was considered by the reflecting part of the community as an

unwarrantable and unjustifiable use of your name – I am happy, Sir, to be enabled to say the author, a person of the name of Locke, is not a native of the United States. He was born in Great Britain. Most true it is, that the vivacity and originality displayed by the author combined with our love for the marvellous, produced a powerful effect. With a few exceptions our whole population, placed implicit confidence in the truth of the statement and would not be convinced of its fallacy, even by the adduction of facts wholly inconsistent with its veracity. Indeed such was the delusion of even intelligent persons, that it has been affirmed a clergyman in the State of Connecticut, on the Sunday subsequent to the appearance of "the Sun" announced to his parishioners the wonderful news, with the comment he had no doubt, some way would soon be devised, of communicating to the Lunarians the blessings of Christianity.'

Herschel, in reply to Paine, stated:

The discoveries in the Moon are highly absurd. One of your countrymen touching here forwarded me the New York paper containing them. At first I was amused – but it was too long, and I could not finish it.—

A letter from Admiral W. H. Smyth to Herschel provides a different viewpoint:

Have you seen the stupid pamphlet published in Paris, purporting to be your observations on the Moon? A copy was presented to the Royal Society. It has a frontispiece of a scene in that luminary, but wholly pointless, a much better [one] might have been made from Munchausen's travels. Paper, types, ink & time, all wasted. Yet a beatled-browed Boeotian questioned me as to its verity!

Had Smyth perused nothing more than the dedication to the King of England contained in the copy of the Paris edition presented to the Royal Society, he would have had just cause to be incensed. The 20-page pamphlet coming from *The Sun* somehow had grown to 160 pages, taking on the appearance of a full-fledged book.

In Paine's letter to Sir John Herschel, a single phrase – 'our love for the marvellous' clearly expresses public expectations of journalism in 1835.

Advances in technology had ushered readers into a new age of discoveries, inventions, and unsurpassed construction. The public was exposed to advances and events which at that period would include:

— the start of experimental photography,
— travel by steamship across oceans and along new canals; steam locomotives moving passengers and freight over expanding networks of railways,
— assembly of a large telescope by Sir John Herschel to study the southern skies,
— the pioneer work of Faraday and Henry, inducing electric currents in wires moved through strong magnetic fields, leading to motors, generators and the distribution of electric power,
— as predicted, the return of Halley's Comet in 1835.

In the euphoria resulting from these developments, was it any wonder the Moon Hoax was accepted without serious question? Why were these readers willing to accept as truth what we today would term impossible?

The public is exposed to advances and events in 1991 which include:

— an extension of experimental photography, using high speed flash and laser techniques,
— air travel by jet, earth orbit by shuttle, and space exploration by probes,
— assembly of the Hubble Telescope to study objects in all parts of the sky,
— pioneering work in wind, solar, nuclear, and fusion power,
— as predicted, the return of Halley's Comet in 1985–1986.

With these advances, would the public accept without question another Moon Hoax? Perhaps closely tied to the launch of Sputnik, the public began to insist on investigative and interpretative reporting. In the latter half of this century, journalists are becoming more responsive to the demands of a reading public with a new level of sophistication and understanding.

In light of this, how should the journalism of Locke be assessed? He wrote very well, created a plausible account and was successful in what he undertook to accomplish. Circulation of *The Sun* increased well beyond his expectations. With about 30 entries in the NUC pre-1956 imprints, his accomplishments cannot be ignored. As the twentieth century draws to a close, hoaxes continue to be perpetrated; such activities are unlikely to disappear in the next. As astronomers and scientists, let us ensure that we do not become innocent victims.

The Star of Bethlehem

WILLIAM FREND

'The stars in their courses fought against Sisera' (Judges, 5.20) and the Israelites were victorious over what had seemed to be overwhelming odds. Stars, not least the planets and comets, whose movements could be observed with the naked eye, played an enormous part in the lives of people in the Ancient World. The American excavators of the Roman frontier fortress and town of Dura Europos (Salilijeh) on the Euphrates were surprised to find that every house they examined had had its own horoscope incised or painted on a wall of one of the rooms. In Rome itself, even Seneca could write (*c.* AD 60) to his friend Marcia, portraying the journey of the soul after death, 'You will see the five planets pursuing their different courses and sparkling down to earth from opposite directions; on their slightest movement hangs the fortunes of nations, and the greatest and smallest happenings are shaped in accordance with the progress of a kindly or unkindly star.' The planets were the lords of time, the 'world rulers' in Paul's description in Galatians (4:3ff), from whose servitude the Christian message had come to preach deliverance. Stars also, however, proclaimed the opposite message of the birth of a deliverer of Israel.

It is not surprising, therefore, that both the birth and death of Jesus of Nazareth would be accompanied by supernatural occurences. 'Darkness at noon' is recorded by Matthew (27:45) at the time of the Crucifixion, and the Star of Bethlehem and the journey thither of the three 'wise men from the East', bearing their gifts to the child Jesus accompany the birth of the Saviour (Matt. 2:1–12), as is told by the same writer.

Should we dismiss all this simply as a legend, or at best an attempt by the Gospel-writer to find events in the Old Testament which he could claim were being fulfilled through those surrounding Jesus' birth? The discussions have gone on since the Reformation, highlighted by Gibbon's ironic account of Matthew's record of astronomical events at the time of the Crucifixion (*Decline and Fall of the Roman Empire*, end ch. XV). Given the historical circumstances of the time, and popular attitudes among the Palestinian Jews, it would appear that while one can find holes in the details of the Birth Narratives, their general truth seems probable.

The main theme of the Lukan birth narrative is the liberation of Israel under a scion of the royal house of David, Jesus. Thus the prophetess Anna in the Temple spoke of Jesus 'to all them that looked for the redemption of Israel in Jerusalem' (Luke 2:38). Before that, Simeon had proclaimed that he was 'set for the fall and rising again of many in Israel' (Luke 2:34), while he himself was awaiting the 'consolation of Israel' (i.e. its liberation from the godless rule of Herod), and the angel who appeared to Mary emphasized that Jesus would be the recipient of the 'throne of his father David' and 'would reign over the house of Jacob for ever' (Luke 1:32–33). The Magnificat (Luke 1: 46–55) is recorded as Mary's expectation of the freedom from injustice and tyranny that her son would bring. The birth of Jesus, therefore, was to be the climax of two generations of messianic expectation, since the capture of Jerusalem by Pompey in 63 BC and the appearance of the Teacher of Righteousness and the Convenanters of Qmran.

Though the idea of a liberator of Israel coming from Galilee sounded slightly ridiculous to some Judæan Jews (John 1:46 and 7:52), Jesus' family was not the run-of-the-mill Galilean settler. Galilee had been conquered by the Jews during the short reign of Aristobulus in 104–103 BC, and soon the area that sixty years before had been 'Galilee of the Gentiles' (1 Macc. 5) had a majority Jewish population, though non-Jews, mainly Syrians, maintained themselves in some of the towns, notably Sepphoris, five miles from Nazareth, and Tiberias. Jesus' family were apparently comparatively recent immigrants. The census of 6 BC, by whosoever called, obliged them to travel south from Nazareth to Bethlehem, where they had to register in Joseph's home near Jerusalem; Bethlehem was also known as 'David's town' (compare Micah 5:2). Joseph, moreover, could trace his ancestry nearly a thousand years back to the royal house of David. It was the sort of lineage we learn from the prophet Zechariah (Zech 12:12) that was carefully preserved.

If Joseph could claim distant royal descent, the family of Mary's cousin Elisabeth could claim ancient priestly connections. Elisabeth herself was a 'daughter of Aaron' (Luke 1:5) and her husband Zacharias was a priest on the staff of the Temple at Jerusalem. There was thus a combination of the royal and priestly heritage in the family that extended from beyond the era of the last liberating dynasty of Israel, the Hasmonaeans, that started one hundred and fifty years before.

In Nazareth, Jesus' family was influential. Even if 'carpenter' does not have a possible meaning of 'learned man', Joseph would have provided most of the technical needs of the community in an agricultural setting. Apart from a holding in Nazareth, the family also seems to have owned land at Capernaum on the Lake of Gennesaret, and mixed there with the local nobility (John 4:46) as well as members of fishing partnerships whence Jesus drew his senior disciples. More significant even for the position they held is John's account of the events at the wedding at Cana (John 2).

Cana was a few miles from Nazareth, and not the same community. Yet when Jesus arrives for the wedding with some disciples (unnamed), he finds his mother in control, and she can instruct servants outside her own immediate neighbourhood to obey her son implicitly, and they did. Again, when Jesus returned to Nazareth after the start of his ministry, and came into the synagogue 'as was his custom', he was immediately offered the scroll to read and comment upon (Luke 4:16ff). He was not yet, however, the 'prophet of Nazareth' that he came later to be known as (Matt. 21:11) and could be criticized severely when he had a good word for Naaman the Syrian; but that he held a leading position in the community is clear. When one adds to this, the profound knowledge of Scripture that his parents must have taught him at an early age (Luke 2:42ff) it is evident that one is dealing with a family which had claims to leadership among the ever-restive Galileans. What is surprising is that the 'hidden years' of Jesus' life lasted so long, and that he hesitated before going down to the Jordan valley where his cousin was conducting a baptism of purification. His future disciples, Peter and Andrew, according to the tradition that John drew on, were already there, and had become John the Baptist's close disciples (John 1:35ff) before he arrived on the scene.

After a generation-long period, Jesus at last emerges as a prophetic and potentially royal (compare John 6:15) figure. Liberation continues to be a major theme in Luke's account of his ministry. The initial careful welcome which his sermon at Nazareth received, resulted from his statement (Luke 4:21) that Israel's prophecy that the Gospel would be preached to the poor and deliverance preached to captives (Is. 61.1) had been fulfilled in the ears of his hearers. Accounts of the births of liberators in Israel's history were carefully kept, with an emphasis on the supernatural. Thus, in the

Book of Judges (ch. 13) Samson is conceived by Manoah's hitherto barren wife, an angel announcing to Manoah the coming miraculous event. Luke's account of the birth of Jesus follows a similar pattern of angelic intervention with the course of nature, and a proclamation of popular hopes of the new era that was to dawn. But while December 25 is an unlikely date for the actual birth of Jesus (shepherds would not be sitting out in the hills with their sheep at this most cold and wet time of the year), there is no reason to disbelieve the main facts of the Lukan story.

Matthew's emphasis is different. Jesus is not only liberator, but is portrayed constantly as a 'greater Moses', proclaiming from a mountain the New Law, just as Moses had proclaimed the Old from Mount Sinai. In this context the symbol of the star becomes all-important; for had not Moses himself written in the Book of Numbers, how Balaam had prophesied the coming of a victorious king of Israel? 'There shall come a star out of Jacob and a sceptre shall rise out of Israel and shall unite the corners of Moab, and destroy all the children of Sheth' (Numbers 24:17). Jesus, as we have seen, was believed to be destined to restore the victorious house of Jacob. In the second great revolt of the Jews against the Romans in AD 132, the leader took the name Bar-Koseba or 'Son of the Star'.

Whatever the Wise Men may have seen, whether a star or a comet, their journey westward from Parthia to Jerusalem about the time of Jesus' birth was remembered and served Matthew's purpose. Herod's reactions suggest an element of truth in the story. Herod had been client-king of Judæa since 37 BC. He was not a Jew, but an Idumæan, from one of the Arab kingdoms conquered by the Israelites at the end of the second century BC. He owed his position to Rome. Before being elevated to the kingship, he had served the Roman cause in repulsing a Parthian invasion of Palestine. In his long reign he had sought to maintain a balance between Jewish and non-Jewish subjects. On the one hand, he began to rebuild the Temple on a grandiose scale, and with the finest cut masonry (the Wailing Wall is simply the massive foundation on which the Temple itself was built). 'Look Teacher, what wonderful stonework and what wonderful buildings' (Mark 13:1) was the reaction of Jesus' contemporaries. But in practically every other way, Herod ruled like a pagan Hellenistic monarch. Hebrew disappeared from his coinage, his capital was the Greek–Syrian city of Cæsarea, not Jerusalem, and another Hellenistic city was built near the site of

Samaria. Above all, Herod was a 'westerner' seeking ultimately to bind the Jews of Palestine with those of the Dispersion in Asia Minor and Greece into a federation loyal to Rome but not completely subject to her. Parthia remained the enemy.

To such a ruler near the end of his life, the announcement of the birth of a Messiah, the King of the Jews, could not have been more unwelcome. Moreover, coming from the east, from Parthia, with the prospect of Parthia's support would have been intolerable. The whole of Herod's lifelong policy would have lain in ruins. No wonder 'he was troubled and all Jerusalem with him' (Matt. 2:3). People knew their Books of Moses and Balaam's prophecy. Herod's tendency toward drastic acts of cruelty was already extending to his own family, and whether the 'massacre of the Innocents' took place as described (Matt. 2:16–17) it was what people believed Herod was capable of doing. The final note in the story of the Flight into Egypt also has an authentic ring. Egypt had provided the refuge through generations of Jews who felt unsafe in their homeland. With Jerusalem and Babylon, Alexandria contained the largest Jewish community in the Ancient World.

The Birth Narratives need to be taken within the context of their times. For two generations there had been Messianic expectations among the Jews, hoping for freedom from Rome, from the oppressive client-rulers whom Rome was sponsoring in Syria and Palestine, and freedom from the extortion of tax-collectors and landowners, whether Jewish or pagan. That some astronomical phenomenon added to the already fervent expectations and hopes for the birth of a new leader seems evident. How it was that these hopes became focused on one particular Galilean family will never be known, but Joseph's family had as good, or better, credentials than most. The carefully recorded details of Jesus' birth in Matthew and Luke should not be dismissed as stories made up by early Christian communities. Instead, they deserve acceptance for their own worth, and also evidence for the aspirations of the Jewish people in the last years of Herod's reign and their hopes that Jesus would be their liberator.

Part III: Miscellaneous

Some Interesting Variable Stars

JOHN ISLES

The following stars are of interest for many reasons. Of course, the periods and ranges of many variables are not constant from one cycle to another.

Star	*R.A.* *h*	*m*	*Declination* *deg.*	*min.*	*Range*	*Type*	*Period days*	*Spectrum*
R Andromedæ	00	24.0	+38	35	5.8–14.9	Mira	409	S
W Andromedæ	02	17.6	+44	18	6.7–14.6	Mira	396	S
U Antliæ	10	35.2	−39	34	5–6	Irregular	–	C
Theta Apodis	14	05.3	−76	48	5–7	Semi-regular	119	M
R Aquarii	23	43.8	−15	17	5.8–12.4	Symbiotic	387	M+Pec
T Aquarii	20	49.9	−05	09	7.2–14.2	Mira	202	M
R Aquilæ	19	06.4	+08	14	5.5–12.0	Mira	284	M
V Aquilæ	19	04.4	−05	41	6.6– 8.4	Semi-regular	353	C
Eta Aquilæ	19	52.5	+01	00	3.5– 4.4	Cepheid	7.2	F–G
U Aræ	17	53.6	−51	41	7.7–14.1	Mira	225	M
R Arietis	02	16.1	+25	03	7.4–13.7	Mira	187	M
U Arietis	03	11.0	+14	48	7.2–15.2	Mira	371	M
R Aurigæ	05	17.3	+53	35	6.7–13.9	Mira	458	M
Epsilon Aurigæ	05	02.0	+43	49	2.9– 3.8	Algol	9892	F+B
R Boötis	14	37.2	+26	44	6.2–13.1	Mira	223	M
W Boötis	14	43.4	+26	32	4.7– 5.4	Semi-regular?	450?	M
X Camelopardalis	04	45.7	+75	06	7.4–14.2	Mira	144	K–M
R Cancri	08	16.6	+11	44	6.1–11.8	Mira	362	M
X Cancri	08	55.4	+17	14	5.6– 7.5	Semi-regular	195?	C
R Canis Majoris	07	19.5	−16	24	5.7– 6.3	Algol	1.1	F
S Canis Minoris	07	32.7	+08	19	6.6–13.2	Mira	333	M
R Canum Ven.	13	49.0	+39	33	6.5–12.9	Mira	329	M
R Carinæ	09	32.2	−62	47	3.9–10.5	Mira	309	M
S Carinæ	10	09.4	−61	33	4.5– 9.9	Mira	149	K–M
l Carinæ	09	45.2	−62	30	3.3– 4.2	Cepheid	35.5	F–K
Eta Carinæ	10	45.1	−59	41	−0.8– 7.9	Irregular	–	Pec
R Cassiopeiæ	23	58.4	+51	24	4.7–13.5	Mira	430	M
S Cassiopeiæ	01	19.7	+72	37	7.9–16.1	Mira	612	S
W Cassiopeiæ	00	54.9	+58	34	7.8–12.5	Mira	406	C
Gamma Cass.	00	56.7	+60	43	1.6– 3.0	Irregular	–	B
Rho Cassiopeiæ	23	54.4	+57	30	4.1– 6.2	Semi-regular	–	F–K
R Centauri	14	16.6	−59	55	5.3–11.8	Mira	546	M
S Centauri	12	24.6	−49	26	7–8	Semi-regular	65	C
T Centauri	13	41.8	−33	36	5.5– 9.0	Semi-regular	90	K–M
S Cephei	21	35.2	+78	37	7.4–12.9	Mira	487	C
T Cephei	21	09.5	+68	29	5.2–11.3	Mira	388	M
Delta Cephei	22	29.2	+58	25	3.5– 4.4	Cepheid	5.4	F–G
Mu Cephei	21	43.5	+58	47	3.4– 5.1	Semi-regular	730	M
U Ceti	02	33.7	−13	09	6.8–13.4	Mira	235	M
W Ceti	00	02.1	−14	41	7.1–14.8	Mira	351	S
Omicron Ceti	02	19.3	−02	59	2.0–10.1	Mira	332	M

Star	*R.A.* *h*	*m*	*Declination* *deg.*	*min.*	*Range*	*Type*	*Period* *days*	*Spectrum*
R Chamælcontis	08	21.8	−76	21	7.5–14.2	Mira	335	M
T Columbæ	05	19.3	−33	42	6.6–12.7	Mira	226	M
R Comæ Ber.	12	04.3	+18	47	7.1–14.6	Mira	363	M
R Coronæ Bor.	15	48.6	+28	09	5.7–14.8	R Coronæ Bor.	–	C
S Coronæ Bor.	15	21.4	+31	22	5.8–14.1	Mira	360	M
T Coronæ Bor.	15	59.6	+25	55	2.0–10.8	Recurr. nova	–	M+Pec
V Coronæ Bor.	15	49.5	+39	34	6.9–12.6	Mira	358	C
W Coronæ Bor.	16	15.4	+37	48	7.8–14.3	Mira	238	M
R Corvi	12	19.6	−19	15	6.7–14.4	Mira	317	M
R Crucis	12	23.6	−61	38	6.4– 7.2	Cepheid	5.8	F–G
R Cygni	19	36.8	+50	12	6.1–14.4	Mira	426	S
U Cygni	20	19.6	+47	54	5.9–12.1	Mira	463	C
W Cygni	21	36.0	+45	22	5.0– 7.6	Semi-regular	131	M
RT Cygni	19	43.6	+48	47	6.0–13.1	Mira	190	M
SS Cygni	21	42.7	+43	35	7.7–12.4	Dwarf nova	50±	K+Pec
CH Cygni	19	24.5	+50	14	5.6– 9.0	Symbiotic	–	M+B
Chi Cygni	19	50.6	+32	55	3.3–14.2	Mira	408	S
R Delphini	20	14.9	+09	05	7.6–13.8	Mira	285	M
U Delphini	20	45.5	+18	05	5.6– 7.5	Semi-regular	110?	M
EU Delphini	20	37.9	+18	16	5.8– 6.9	Semi-regular	60	M
Beta Doradus	05	33.6	−62	29	3.5– 4.1	Cepheid	9.8	F–G
R Draconis	16	32.7	+66	45	6.7–13.2	Mira	246	M
T Eridani	03	55.2	−24	02	7.2–13.2	Mira	252	M
R Fornacis	02	29.3	−26	06	7.5–13.0	Mira	389	C
R Geminorum	07	07.4	+22	42	6.0–14.0	Mira	370	S
U Geminorum	07	55.1	+22	00	8.2–14.9	Dwarf nova	105±	Pec+M
Zeta Geminorum	07	04.1	+20	34	3.6– 4.2	Cepheid	10.2	F–G
Eta Geminorum	06	14.9	+22	30	3.2– 3.9	Semi-regular	233	M
S Gruis	22	26.1	−48	26	6.0–15.0	Mira	402	M
S Herculis	16	51.9	+14	56	6.4–13.8	Mira	307	M
U Herculis	16	25.8	+18	54	6.4–13.4	Mira	406	M
Alpha Herculis	17	14.6	+14	23	2.7– 4.0	Semi-regular	–	M
68, u Herculis	17	17.3	+33	06	4.7– 5.4	Algol	2.1	B+B
R Horologii	02	53.9	−49	53	4.7–14.3	Mira	408	M
U Horologii	03	52.8	−45	50	6–14	Mira	348	M
R Hydræ	13	29.7	−23	17	3.5–10.9	Mira	389	M
U Hydræ	10	37.6	−13	23	4.3– 6.5	Semi-regular	450?	C
VW Hydri	04	09.1	−71	18	8.4–14.4	Dwarf nova	27±	Pec
R Leonis	09	47.6	+11	26	4.4–11.3	Mira	310	M
R Leonis Minoris	09	45.6	+34	31	6.3–13.2	Mira	372	M
R Leporis	04	59.6	−14	48	5.5–11.7	Mira	427	C
Y Libræ	15	11.7	−06	01	7.6–14.7	Mira	276	M
RS Libræ	15	24.3	−22	55	7.0–13.0	Mira	218	M
Delta Libræ	15	01.0	−08	31	4.9– 5.9	Algol	2.3	A
R Lyncis	07	01.3	+55	20	7.2–14.3	Mira	379	S
R Lyræ	18	55.3	+43	57	3.9– 5.0	Semi-regular	46?	M
RR Lyræ	19	25.5	+42	47	7.1– 8.1	RR Lyræ	0.6	A–F
Beta Lyræ	18	50.1	+33	22	3.3– 4.4	Eclipsing	12.9	B
U Microscopii	20	29.2	−40	25	7.0–14.4	Mira	334	M
U Monocerotis	07	30.8	−09	47	5.9– 7.8	RV Tauri	91	F–K
V Monocerotis	06	22.7	−02	12	6.0–13.9	Mira	340	M
R Normæ	15	36.0	−49	30	6.5–13.9	Mira	508	M
T Normæ	15	44.1	−54	59	6.2–13.6	Mira	241	M
R Octantis	05	26.1	−86	23	6.3–13.2	Mira	405	M
S Octantis	18	08.7	−86	48	7.2–14.0	Mira	259	M
V Ophiuchi	16	26.7	−12	26	7.3–11.6	Mira	297	C
X Ophiuchi	18	38.3	+08	50	5.9– 9.2	Mira	329	M
RS Ophiuchi	17	50.2	−06	43	4.3–12.5	Recurr. nova	–	OB+M
U Orionis	05	55.8	+20	10	4.8–13.0	Mira	368	M
W Orionis	05	05.4	+01	11	5.9– 7.7	Semi-regular	212	C

Star	*R.A.* *h*	*m*	*Declination* *deg.*	*min.*	*Range*	*Type*	*Period* *days*	*Spectrum*
Alpha Orionis	05	55.2	+07	24	0.0– 1.3	Semi-regular	2335	M
S Pavonis	19	55.2	−59	12	6.6–10.4	Semi-regular	381	M
Kappa Pavonis	18	56.9	−67	14	3.9– 4.8	Cepheid	9.1	G
R Pegasi	23	06.8	+10	33	6.9–13.8	Mira	378	M
Beta Pegasi	23	03.8	+28	05	2.3– 2.7	Irregular	–	M
X Persei	03	55.4	+31	03	6.0– 7.0	Gamma Cass.	–	O9.5
Beta Persei	03	08.2	+40	57	2.1– 3.4	Algol	2.9	B
Rho Persei	03	05.2	+38	50	3.3– 4.0	Semi-regular	50?	M
Zeta Phœnicis	01	08.4	−55	15	3.9– 4.4	Algol	1.7	B+B
R Pictoris	04	46.2	−49	15	6.4–10.1	Semi-regular	171	M
L² Puppis	07	13.5	−44	39	2.6– 6.2	Semi-regular	141	M
T Pyxidis	09	04.7	−32	23	6.5–15.3	Recurr. nova	7000±	Pec
U Sagittæ	19	18.8	+19	37	6.5– 9.3	Algol	3.4	B+G
WZ Sagittæ	20	07.6	+17	42	7.0–15.5	Dwarf nova	11900±	A
R Sagittarii	19	16.7	−19	18	6.7–12.8	Mira	270	M
RR Sagittarii	19	55.9	−29	11	5.4–14.0	Mira	336	M
RT Sagittarii	20	17.7	−39	07	6.0–14.1	Mira	306	M
RU Sagittarii	19	58.7	−41	51	6.0–13.8	Mira	240	M
RY Sagittarii	19	16.5	−33	31	5.8–14.0	R Coronæ Bor.	–	G
RR Scorpii	16	56.6	−30	35	5.0–12.4	Mira	281	M
RS Scorpii	16	55.6	−45	06	6.2–13.0	Mira	320	M
RT Scorpii	17	03.5	−36	55	7.0–15.2	Mira	449	S
S Sculptoris	00	15.4	−32	03	5.5–13.6	Mira	363	M
R Scuti	18	47.5	−05	42	4.2– 8.6	RV Tauri	146	G–K
R Serpentis	15	50.7	+15	08	5.2–14.4	Mira	356	M
S Serpentis	15	21.7	+14	19	7.0–14.1	Mira	372	M
T Tauri	04	22.0	+19	32	9.3–13.5	Irregular	–	F–K
SU Tauri	05	49.1	+19	04	9.1–16.9	R Coronæ Bor.	–	G
Lambda Tauri	04	00.7	+12	29	3.4– 3.9	Algol	4.0	B+A
R Trianguli	02	37.0	+34	16	5.4–12.6	Mira	267	M
R Ursæ Majoris	10	44.6	+68	47	6.5–13.7	Mira	302	M
T Ursæ Majoris	12	36.4	+59	29	6.6–13.5	Mira	257	M
U Ursæ Minoris	14	17.3	+66	48	7.1–13.0	Mira	331	M
R Virginis	12	38.5	+06	59	6.1–12.1	Mira	146	M
S Virginis	13	33.0	−07	12	6.3–13.2	Mira	375	M
SS Virginis	12	25.3	+00	48	6.0– 9.6	Semi-regular	364	C
R Vulpeculæ	21	04.4	+23	49	7.0–14.3	Mira	137	M
Z Vulpeculæ	19	21.7	+25	34	7.3– 8.9	Algol	2.5	B+A

Mira Stars: maxima, 1996

JOHN ISLES

Below are given predicted dates of maxima for Mira stars that reach magnitude 7.5 or brighter at an average maximum. Individual maxima can in some cases be brighter or fainter than average by a magnitude or more, and all dates are only approximate. The positions, extreme ranges and mean periods of these stars can all be found in the preceding list of interesting variable stars.

Star	*Mean magnitude at maximum*	*Dates of maxima*
W Andromedæ	7.4	Apr. 1
R Aquarii	6.5	Feb. 7
R Aquilæ	6.1	Jan. 18, Oct. 30
R Boötis	7.2	Mar. 22, Oct. 31
R Cancri	6.8	Oct. 30
S Canis Minoris	7.5	June 30
R Carinæ	4.6	Jan. 18, Nov. 23
S Carinæ	5.7	Apr. 8, Sep. 5
R Cassiopeiæ	7.0	Nov. 3
R Centauri	5.8	June 21
T Cephei	6.0	Dec. 29
U Ceti	7.5	Mar. 5, Oct. 25
Omicron Ceti	3.4	May 16
T Columbæ	7.5	Feb. 12, Sep. 26
S Coronæ Borealis	7.3	Oct. 11
V Coronæ Borealis	7.5	May 11
R Corvi	7.5	Mar. 23
U Cygni	7.2	Oct. 4
RT Cygni	7.3	June 21, Dec. 28
Chi Cygni	5.2	May 31
R Geminorum	7.1	Oct. 21
U Herculis	7.5	Apr. 28
R Horologii	6.0	July 4
U Horologii	7	Mar. 1
R Hydræ	4.5	Jan. 30
R Leonis	5.8	May 12
R Leonis Minoris	7.1	Aug. 7
R Leporis	6.8	Apr. 27
RS Libræ	7.5	July 26
V Monocerotis	7.0	Oct. 26
T Normæ	7.4	Apr. 26, Dec. 23

Star	*Mean magnitude at maximum*	*Dates of maxima*
V Ophiuchi	7.5	July 24
X Ophiuchi	6.8	Feb. 6
U Orionis	6.3	Nov. 17
R Sagittarii	7.3	June 29
RR Sagittarii	6.8	May 3
RT Sagittarii	7.0	Mar. 30
RU Sagittarii	7.2	Jan. 7, Sep. 4
RR Scorpii	5.9	Sep. 22
RS Scorpii	7.0	Apr. 15
S Sculptoris	6.7	Nov. 29
R Serpentis	6.9	Mar. 25
R Trianguli	6.2	May 28
R Ursæ Majoris	7.5	May 24
R Virginis	6.9	Jan. 14, June 9, Nov. 1
S Virginis	7.0	Aug. 29

Some Interesting Double Stars

R. W. ARGYLE

The positions given below correspond to epoch 1996.0

Name	*Magnitudes*	*Separation in seconds of arc*	*Position angle, degrees*	*Notes*
Gamma Andromedæ	2.3, 5.0	9.4	064	Yellow, blue. B is again double.
Zeta Aquarii	4.3, 4.5	2.0	196	Slowly widening.
53 Aquarii	6.4, 6.6	2.0	353	Long period binary. Closing.
Gamma Arietis	4.8, 4.8	7.6	000	Very easy. Both white.
Epsilon Arietis	5.2, 5.5	1.5	208	Binary. Both white.
Theta Aurigæ	2.6, 7.1	3.7	310	Stiff test for 3-in.
44 Boötis	5.3, 6.2	2.0	051	Period 246 years.
Xi Boötis	4.7, 7.0	6.9	322	Fine contrast. Easy.
Epsilon Boötis	2.5, 4.9	2.8	342	Yellow, blue. Fine pair.
Zeta Cancri	5.6, 6.2	6.0	074	A again double.
Iota Cancri	4.2, 6.6	30.4	307	Easy. Yellow, blue.
Alpha Canum Ven.	2.9, 5.5	19.6	228	Easy. Yellow, bluish.
Upsilon Carinæ	3.1, 6.1	5.0	127	Fixed.
Eta Cassiopeiæ	3.4, 7.5	12.7	315	Easy. Creamy, bluish.
Alpha Centauri	0.0, 1.2	17.3	218	Very easy. Period 80 years. Closing.
Gamma Centauri	2.9, 2.9	1.2	351	Period 84 years. Closing. Both yellow.
3 Centauri	4.5, 6.0	7.8	105	Both white.
Beta Cephei	3.2, 7.9	13.4	250	Easy with a 3-in.
Delta Cephei	var, 7.5	41.3	192	Very easy.
Xi Cephei	4.4, 6.5	8.0	275	White, blue.
Gamma Ceti	3.5, 7.3	2.9	294	Not too easy.
Alpha Circini	3.2, 8.6	15.7	230	PA slowly decreasing.
Zeta Coronæ Bor.	5.1, 6.0	6.3	305	PA slowly increasing.
Delta Corvi	3.0, 9.2	24	214	Easy with a 3-in.
Alpha Crucis	1.4, 1.9	4.2	114	Third star in a low-power field.
Mu Crucis	4.3, 5.3	34.9	017	Fixed. Both white.
Beta Cygni	3.1, 5.1	34.1	054	Glorious. Yellow, blue.
61 Cygni	5.2, 6.0	30.3	149	Nearby binary. Period 722 years.
Gamma Delphini	4.5, 5.5	9.3	267	Easy. Yellowish, greenish.
Epsilon Draconis	3.8, 7.4	3.2	019	Slow binary.
Nu Draconis	4.9, 4.9	61.7	311	Binocular pair.
f Eridani	4.8, 5.3	8.2	215	Pale yellow.
p Eridani	5.8, 5.8	11.5	192	Period 483 years.
Theta Eridani	3.4, 4.5	8.3	090	Both white.

Name	Magnitudes	Separation in seconds of arc	Position angle, degrees	Notes
Alpha Geminorum	1.9, 2.9	3.5	072	Widening. Easy with a 3-in.
Delta Geminorum	3.5, 8.2	5.8	224	Not too easy.
Alpha Herculis	var, 5.4	4.6	106	Red, green. Binary.
Delta Herculis	3.1, 8.2	10.5	277	Optical pair. Distance increasing.
Zeta Herculis	2.9, 5.5	1.4	054	Fine, rapid binary. Period 34 years.
Epsilon Hydræ	3.3, 6.8	2.7	298	PA slowly increasing.
Theta Indi	4.5, 7.0	6.7	266	Fine contrast.
Gamma Leonis	2.2, 3.5	4.4	125	Binary, 619 years.
Pi Lupi	4.6, 4.7	1.7	065	Widening.
Alpha Lyræ	0.0, 9.5	76	182	Optical pair. B is faint.
Epsilon[1] Lyræ	5.0, 6.1	2.6	351	Quadruple system. Both
Epsilon[2] Lyræ	5.2, 5.5	2.3	084	pairs visible in a 3-in.
Zeta Lyræ	4.3, 5.9	44.0	150	Fixed. Easy double.
Beta Muscæ	3.7, 4.0	1.3	040	Both white. Closing.
70 Ophiuchi	4.2, 6.0	2.6	164	Opening. Easy in 3-in.
Beta Orionis	0.1, 6.8	9.5	202	Can be seen with 3-in.
Iota Orionis	2.8, 6.9	11.8	141	Enmeshed in nebulosity.
Theta Orionis	6.7, 7.9	8.7	032	Trapezium in M42.
	5.1, 6.7	13.4	061	
Sigma Orionis	4.0, 10.3	11.4	238	Quintuple. A is a
	6.5, 7.5	30.1	231	close double.
Zeta Orionis	1.9, 4.0	2.4	162	Can be split in 3-in.
Xi Pavonis	4.4, 8.6	3.3	155	Orange and white.
Eta Persei	3.8, 8.5	28.5	300	Yellow, bluish.
Beta Phœnicis	4.0, 4.2	1.5	324	Slowly widening.
Beta Piscis Aust.	4.4, 7.9	30.4	172	Optical pair. Fixed.
Alpha Piscium	4.2, 5.1	1.8	274	Binary, 933 years.
Kappa Puppis	4.5, 4.7	9.8	318	Both white.
Alpha Scorpii	1.2, 5.4	2.7	274	Red, green. Difficult.
Nu Scorpii	4.3, 6.4	41.2	337	Both again double.
Theta Serpentis	4.5, 5.4	22.3	103	Fixed. Very easy.
Alpha Tauri	0.9, 11.1	133	031	Wide, but B very faint in small telescopes.
Iota Trianguli	5.3, 6.9	3.9	070	Slow binary.
Beta Tucanæ	4.4, 4.8	27.1	170	Both again double.
Delta Tucanæ	4.5, 9.0	6.9	282	White, reddish.
Zeta Ursæ Majoris	2.3, 4.0	14.4	151	Very easy. Naked eye pair with Alcor.
Xi Ursæ Majoris	4.3, 4.8	1.3	304	Binary, 60 years. Closing. Needs a 4-in.
Delta Velorum	2.1, 5.1	2.0	140	Slowly closing.
s Velorum	6.2, 6.5	13.5	218	Fixed.
Gamma Virginis	3.5, 3.5	2.1	274	Binary, 168 years. Closing.
Theta Virginis	4.4, 9.4	7.1	343	Not too easy.
Gamma Volantis	3.9, 5.8	13.8	299	Very slow binary.

Some Interesting Nebulæ and Clusters

Object	R.A.		Dec.		*Remarks*
	h	*m*			
M.31 Andromedæ	00	40.7	+41	05	Great Galaxy, visible to naked eye.
H.VIII 78 Cassiopeiæ	00	41.3	+61	36	Fine cluster, between Gamma and Kappa Cassiopeiæ.
M.33 Trianguli	01	31.8	+30	28	Spiral. Difficult with small apertures.
H.VI 33–4 Persei	02	18.3	+56	59	Double cluster; Sword-handle.
△142 Doradûs	05	39.1	−69	09	Looped nebula round 30 Doradûs. Naked-eye. In Large Cloud of Magellan.
M.1 Tauri	05	32.3	+22	00	Crab Nebula, near Zeta Tauri.
M.42 Orionis	05	33.4	−05	24	Great Nebula. Contains the famous Trapezium, Theta Orionis.
M.35 Geminorum	06	06.5	+24	21	Open cluster near Eta Geminorum.
H.VII 2 Monocerotis	06	30.7	+04	53	Open cluster, just visible to naked eye.
M.41 Canis Majoris	06	45.5	−20	42	Open cluster, just visible to naked eye.
M.47 Puppis	07	34.3	−14	22	Mag. 5,2. Loose cluster.
H.IV 64 Puppis	07	39.6	−18	05	Bright planetary in rich neighbourhood.
M.46 Puppis	07	39.5	−14	42	Open cluster.
M.44 Cancri	08	38	+20	07	Præsepe. Open cluster near Delta Cancri. Visible to naked eye.
M.97 Ursæ Majoris	11	12.6	+55	13	Owl Nebula, diameter 3′. Planetary.
Kappa Crucis	12	50.7	−60	05	'Jewel Box'; open cluster, with stars of contrasting colours.
M.3 Can. Ven.	13	40.6	+28	34	Bright globular.
Omcga Centauri	13	23.7	−47	03	Finest of all globulars. Easy with naked eye.
M.80 Scorpii	16	14.9	−22	53	Globular, between Antares and Beta Scorpionis.
M.4 Scorpii	16	21.5	−26	26	Open cluster close to Antares.
M.13 Herculis	16	40	+36	31	Globular. Just visible to naked eye.
M.92 Herculis	16	16.1	+43	11	Globular. Between Iota and Eta Herculis.
M.6 Scorpii	17	36.8	−32	11	Open cluster; naked eye.
M.7 Scorpii	17	50.6	−34	48	Very bright open cluster; naked eye.
M.23 Sagittarii	17	54.8	−19	01	Open cluster nearly 50′ in diameter.
H.IV 37 Draconis	17	58.6	+66	38	Bright planetary.
M.8 Sagittarii	18	01.4	−24	23	Lagoon Nebula. Gaseous. Just visible with naked eye.
NGC 6572 Ophiuchi	18	10.9	+06	50	Bright planetary, between Beta Ophiuchi and Zeta Aquilæ.
M.17 Sagittarii	18	18.8	−16	12	Omega Nebula. Gaseous. Large and bright.
M.11 Scuti	18	49.0	−06	19	Wild Duck. Bright open cluster.
M.57 Lyræ	18	52.6	+32	59	Ring Nebula. Brightest of planetaries.
M.27 Vulpeculæ	19	58.1	+22	37	Dumb-bell Nebula, near Gamma Sagittæ.
H.IV 1 Aquarii	21	02.1	−11	31	Bright planetary near Nu Aquarii.
M.15 Pegasi	21	28.3	+12	01	Bright globular, near Epsilon Pegasi.
M.39 Cygni	21	31.0	+48	17	Open cluster between Deneb and Alpha Lacertæ. Well seen with low powers.

Our Contributors

Dr Fred Watson has been carrying out his researches at the Royal Greenwich Observatory; possibly his greatest contribution to date has been his pioneer work in fibre optics. He has now moved to Australia, to take charge of the UK Schmidt Telescope at Siding Spring.

Dr Paul Murdin, O.B.E., is one of the world's leading astronomers; he has been Director of the Royal Observatory Edinburgh, was for a time in charge of the British telescopes in La Palma, and is now at P.P.A.R.C. In addition to his technical work, he is well known for his popular books and broadcasts.

Michael De Faubert Maunder is a research chemist who has a lifelong interest in astronomy; he is a specialist in astronomical photography, and is an avid 'eclipse chaser'.

Dame Kathleen Ollerenshaw, D.B.E., D.STJ.D.L., was born in Manchester in 1912; she graduated in mathematics from Oxford in 1934, and is by profession a (pure) mathematician. She is a former Lord Mayor of Manchester and a Freeman of the City; she has been awarded honorary doctorates at five universities, and in 1971 was awarded the D.B.E. for 'services to education'. She continues her mathematical work, and has just completed a major piece of research. At the age of 78 she took up astronomy for the first time, and is finding it of enthralling interest, now achieving, despite limited opportunity, spectacular results with CCD imaging.

Dr Duncan Steel is a specialist in planetary dynamics. He works at the Anglo-Australian Observatory at Coonabarabran, in New South Wales, and at the University of Adelaide.

Professor Shaun Hughes joined the H.S.T. Extragalactic Distance Scale Key Project in 1992, working at the California Institute of Technology with Jeremy Mould. Now at the Royal Greenwich Observatory, he is still able to spend about half his time on the Key Project, and as well as analysing H.S.T. images has been using the

UK telescopes on La Palma to obtain photometry of target galaxies for ground-based calibrations of the H.S.T. measurements.

Richard McKim, a schoolmaster by profession, is a well-known amateur astronomer who specializes in planetary observation and historical research. In 1993–5 he served as President of the British Astronomical Association, and is the Director of its Mars Section.

Professor J. E. Kennedy has had a long research career in physics, and is now Professor Emeritus of Physics at the University of Saskatchewan, at Saskatoon in Canada.

The Rev. Professor William Frend is Professor Emeritus of Ecclesiastical History at the University of Glasgow, and sometime Fellow of Gonville and Caius College, Cambridge. After retirement from Glasgow he spent six years as Priest-in-Charge of the Barnwell group of parishes in Northamptonshire. Since 1990 he has lived near Cambridge. He is a Fellow of the British Academy, the Royal Society of Edinburgh, and the Society of Antiquities of London.

Astronomical Societies in Great Britain

British Astronomical Association
Assistant Secretary: Burlington House, Piccadilly, London W1V 9AG.
Meetings: Lecture Hall of Scientific Societies, Civil Service Commission Building, 23 Savile Row, London W1. Last Wednesday each month (Oct.–June). 1700 hrs and some Saturday afternoons.

Association for Astronomy Education
Secretary: Bob Kibble, 34 Ackland Crescent, Denmark Hill, London SE5 8EQ.

Astronomy Ireland
Secretary: Tony Ryan, PO Box 2888, Dublin 1. Tel. 01-459-88-83.
Meetings: 2nd and 4th Mondays of each month. Telescope meetings, every clear Saturday.

Federation of Astronomical Societies
Secretary: Mrs Christine Sheldon, Whitehaven, Lower Moor, Pershore, Worcs.

Junior Astronomical Society of Ireland
Secretary: K. Nolan, 5 St Patrick's Crescent, Rathcoole, Co. Dublin.
Meetings: The Royal Dublin Society, Ballsbridge, Dublin 4. Monthly.

Aberdeen and District Astronomical Society
Secretary: Stephen Graham, 25 Davidson Place, Northfield, Aberdeen.
Meetings: Robert Gordon's Institute of Technology, St Andrew's Street, Aberdeen. Friday 7.30 p.m.

Altrincham and District Astronomical Society
Secretary: Colin Henshaw, 10 Delamore Road, Gatley, Cheadle, Cheshire.
Meetings: Public Library, Timperley. 1st Friday of each month, 7.30 p.m.

Astra Astronomy Section
Secretary: Ian Downie, 151 Sword Street, Glasgow G31.
Meetings: Public Library, Airdrie. Weekly.

Aylesbury Astronomical Society
Secretary: Nigel Sheridan, 22 Moor Park, Wendover, Bucks.
Meetings: 1st Monday in month. Details from Secretary.

Bassetlaw Astronomical Society
Secretary: H. Moulson, 5 Magnolia Close, South Anston, South Yorks.
Meetings: Rhodesia Village Hall, Rhodesia, Worksop, Notts. On 2nd and 4th Tuesdays of month at 8 p.m.

Batley & Spenborough Astronomical Society
Secretary: Robert Morton, 22 Links Avenue, Cleckheaton, West Yorks BD19 4EG.
Meetings: Milner K. Ford Observatory, Wilton Park, Batley. Every Thursday, 7.30 p.m.

Bedford Astronomical Society
Secretary: D. Eagle, 24 Copthorne Close, Oakley, Bedford.
Meetings: Bedford School, Burnaby Rd, Bedford. Last Tuesday each month.

Bingham & Brookes Space Organization
Secretary: N. Bingham, 15 Hickmore's Lane, Lindfield, W. Sussex.

Birmingham Astronomical Society
Secretary: J. Spittles, 28 Milverton Road, Knowle, Solihull, West Midlands.
Meetings: Room 146, Aston University, last Tuesday each month, Sept. to June (except December moved to 1st week in January).

Blackpool & District Astronomical Society
Secretary: J. L. Crossley, 24 Fernleigh Close, Bispham, Blackpool, Lancs.

Bolton Astronomical Society
Secretary: Peter Miskiw, 9 Hedley Street, Bolton.

Border Astronomical Society
Secretary: David Pettit, 14 Shap Grove, Carlisle, Cumbria.

Boston Astronomers
Secretary: B. Tongue, South View, Fen Road, Stickford, Boston.
Meetings: Details from the Secretary.

Bradford Astronomical Society
Secretary: John Schofield, Briar Lea, Bromley Road, Bingley, W. Yorks.
Meetings: Eccleshill Library, Bradford 2. Monday fortnightly (with occasional variations).

Braintree, Halstead & District Astronomical Society
Secretary: Heather Reeder, The Knoll, St Peters in the Field, Braintree, Essex.
Meetings: St Peter's Church Hall, St Peter's Road, Braintree, Essex. 3rd Thursday each month, 8 p.m.

Bridgend Astronomical Society
Secretary: Clive Down, 10 Glan y Llyn, Broadlands, North Cornelly, Bridgend.
Meetings: G.P. Room, Recreation Centre, Bridgend, 1st and 3rd Friday monthly, 7.30 p.m.

Bridgwater Astronomical Society
Secretary: W. L. Buckland, 104 Polden Street, Bridgwater, Somerset.
Meetings: Room D10, Bridgwater College, Bath Road Centre, Bridgwater. 2nd Wednesday each month, Sept.–June.

Brighton Astronomical Society
Secretary: Mrs B. C. Smith, Flat 2, 23 Albany Villas, Hove, Sussex BN3 2RS.
Meetings: Preston Tennis Club, Preston Drive, Brighton. Weekly, Tuesdays.

Bristol Astronomical Society
Secretary: Geoff Cane, 9 Sandringham Road, Stoke Gifford, Bristol.
Meetings: Royal Fort (Rm G44), Bristol University. Every Friday each month, Sept.–May. Fortnightly, June–August.

Cambridge Astronomical Association
Secretary: R. J. Greening, 20 Cotts Croft, Great Chishill, Royston, Herts.
Meetings: Venues as published in newsletter. 1st and 3rd Friday each month, 8 p.m.

Cardiff Astronomical Society
Secretary: D. W. S. Powell, 1 Tal-y-Bont Road, Ely, Cardiff.
Meeting Place: Room 230, Dept. Law, University College, Museum Avenue, Cardiff. Alternate Thursdays, 8 p.m.

Castle Point Astronomy Club
Secretary: Miss Zena White, 43 Lambeth Road, Eastwood, Essex.
Meetings: St Michael's Church, Thundersley. Most Wednesdays, 8 p.m.

Chelmsford Astronomers
Secretary: Brendan Clark, 5 Borda Close, Chelmsford, Essex.
Meetings: Once a month.

Chelmsford and District Astronomical Society
Secretary: Miss C. C. Puddick, 6 Walpole Walk, Rayleigh, Essex.
Meetings: Sandon House School, Sandon, near Chelmsford. 2nd and last Monday of month. 7.45 p.m.

Chester Astronomical Society
Secretary: Mrs S. Brooks, 39 Halton Road, Great Sutton, South Wirral.
Meetings: Southview Community Centre, Southview Road, Chester. Last Monday each month except Aug. and Dec., 7.30 p.m.

Chester Society of Natural Science Literature and Art
Secretary: Paul Braid, 'White Wing', 38 Bryn Avenue, Old Colwyn, Colwyn Bay, Clwyd.
Meetings: Grosvenor Museum, Chester. Fortnightly.

Chesterfield Astronomical Society
Secretary: P. Lisewski, 148 Old Hall Road, Brampton, Chesterfield.
Meetings: Barnet Observatory, Newbold. Each Friday.

Clacton & District Astronomical Society
Secretary: C. L. Haskell, 105 London Road, Clacton-on-Sea, Essex.

Cleethorpes & District Astronomical Society
Secretary: C. Illingworth, 38 Shaw Drive, Grimsby, S. Humberside.
Meetings: Beacon Hill Observatory, Cleethorpes. 1st Wednesday each month.

Cleveland & Darlington Astronomical Society
Secretary: Neil Haggath, 5 Fountains Crescent, Eston, Middlesbrough, Cleveland.
Meetings: Elmwood Community Centre, Greens Lane, Hartburn, Stockton-on-Tees. Monthly, usually 2nd Friday.

Colchester Amateur Astronomers
Secretary: F. Kelly, 'Middleton', Church Road, Elmstead Market, Colchester, Essex.
Meetings: William Loveless Hall, High Street, Wivenhoe. Friday evenings. Fortnightly.

Cotswold Astronomical Society
Secretary: Trevor Talbot, Innisfree, Winchcombe Road, Sedgebarrow, Worcs.
Meetings: Fortnightly in Cheltenham or Gloucester.

Coventry & Warwicks Astronomical Society
Secretary: V. Cooper, 5 Gisburn Close, Woodloes Park, Warwick.
Meetings: Coventry Technical College. 1st Friday each month, Sept.–June.

Crawley Astronomical Society
Secretary: G. Cowley, 67 Climpixy Road, Ifield, Crawley, Sussex.
Meetings: Crawley College of Further Education. Monthly Oct.–June.

Crayford Manor House Astronomical Society
Secretary: R. H. Chambers, Manor House Centre, Crayford, Kent.
Meetings: Manor House Centre, Crayford. Monthly during term-time.

Croydon Astronomical Society
Secretary: Simon Bailey, 39 Sanderstead Road, South Croydon, Surrey.
Meetings: Lecture Theatre, Royal Russell School, Combe Lane, South Croydon. Alternate Fridays, 7.45 p.m.

Derby & District Astronomical Society
Secretary: Jane D. Kirk, 7 Cromwell Avenue, Findern, Derby.
Meetings: At home of Secretary. 1st and 3rd Friday each month, 7.30 p.m.

Doncaster Astronomical Society
Secretary: J. A. Day, 297 Lonsdale Avenue, Intake, Doncaster.
Meetings: Fridays, weekly.

Dundee Astronomical Society
Secretary: G. Young, 37 Polepark Road, Dundee, Angus.
Meetings: Mills Observatory, Balgay Park, Dundee. 1st Friday each month, 7.30 p.m. Sept.–April.

Easington and District Astronomical Society
Secretary: T. Bradley, 52 Jameson Road, Hartlepool, Co. Durham.
Meetings: Easington Comprehensive School, Easington Colliery. Every 3rd Thursday throughout the year, 7.30 p.m.

Eastbourne Astronomical Society
Secretary: D. C. Gates, Apple Tree Cottage, Stunts Green, Hertsmonceux, East Sussex.
Meetings: St Aiden's Church Hall, 1 Whitley Road, Eastbourne. Monthly (except July and August).

East Lancashire Astronomical Society
Secretary: D. Chadwick, 16 Worston Lane, Great Harwood, Blackburn BB6 7TH.
Meetings: As arranged. Monthly.

Astronomical Society of Edinburgh
Secretary: R. G. Fenoulhet, 7 Greenend Gardens, Edinburgh EH17 7QB.
Meetings: City Observatory, Calton Hill, Edinburgh. Monthly.

Edinburgh University Astronomical Society
Secretary: c/o Dept. of Astronomy, Royal Observatory, Blackford Hill, Edinburgh.

Ewell Astronomical Society
Secretary: Edward Hanna, 91 Tennyson Avenue, Motspur Park, Surrey.
Meetings: 1st Friday of each month.

Exeter Astronomical Society
Secretary: Miss J. Corey, 5 Egham Avenue, Topsham Road, Exeter.
Meetings: The Meeting Room, Wynards, Magdalen Street, Exeter. 1st Thursday of month.

Farnham Astronomical Society
Secretary: Laurence Anslow, 14 Wellington Lane, Farnham, Surrey.
Meetings: Church House, Union Road, Farnham. 2nd Monday each month, 7.45 p.m.

Fitzharry's Astronomical Society (Oxford & District)
Secretary: Mark Harman, 20 Lapwing Lane, Cholsey, Oxon.
Meetings: All Saints Methodist Church, Dorchester Crescent, Abingdon, Oxon.

Forest Astronomical Society
Chairman: Tony Beale, 8 Mill Lane, Lower Beeding, West Sussex.
Meetings: 1st Wednesday each month, juniors following Fridays. For location contact chairman.

Furness Astronomical Society
Secretary: A. Thompson, 52 Ocean Road, Walney Island, Barrow-in-Furness, Cumbria.
Meetings: St Mary's Church Centre, Dalton-in-Furness. 2nd Saturday in month, 7.30 p.m. No August meeting.

Fylde Astronomical Society
Secretary: 28 Belvedere Road, Thornton, Lancs.
Meetings: Stanley Hall, Rossendale Avenue South. 1st Wednesday each month.

Astronomical Society of Glasgow
Secretary: Malcolm Kennedy, 32 Cedar Road, Cumbernauld, Glasgow.
Meetings: University of Strathclyde, George St., Glasgow. 3rd Thursday each month, Sept.–April.

Great Ellingham and District Astronomy Club
Secretary: Andrew Briggs, Avondale, Norwich Road, Besthorpe, Norwich.
Meetings: Great Ellingham Recreation Centre, Watton Road, Great Ellingham. 2nd or 3rd Friday each month (check with Secretary), 7.15 p.m.

Greenock Astronomical Society
Secretary: Carl Hempsey, 49 Brisbane Street, Greenock.
Meetings: Greenock Arts Guild, 3 Campbell Street, Greenock.

Grimsby Astronomical Society
Secretary: R. Williams, 14 Richmond Close, Grimsby, South Humberside.
Meetings: Secretary's home. 2nd Thursday each month, 7.30 p.m.

Guernsey: La Société Guernesiaise Astronomy Section
Secretary: G. Falla, Highcliffe, Avenue Beauvais, Ville du Roi, St Peter's Port, Guernsey.
Meetings: The Observatory, St Peter's, Tuesdays, 8 p.m.

Guildford Astronomical Society
Secretary: A. Langmaid, 22 West Mount, Guildford, Surrey.
Meetings: Guildford Institute, Ward Street, Guildford. 1st Thursday each month, except July and August, 7.30 p.m.

Gwynedd Astronomical Society
Secretary: P. J. Curtis, Ael-y-bryn, Malltraeth St Newborough, Anglesey, Gwynedd.
Meetings: Physics Lecture Room, Bangor University. 1st Thursday each month, 7.30 p.m.

The Hampshire Astronomical Group
Secretary: R. F. Dodd, 1 Conifer Close, Cowplain, Waterlooville, Hants.
Meetings: Clanfield Observatory. Each Friday, 7.30 p.m.

Astronomical Society of Haringey
Secretary: Wally Baker, 58 Stirling Road, Wood Green, London N22.
Meetings: The Hall of the Good Shepherd, Berwick Road, Wood Green. 3rd Wednesday each month, 8 p.m.

Harrogate Astronomical Society
Secretary: P. Barton, 31 Gordon Avenue, Harrogate, North Yorkshire.
Meetings: Harlow Hill Methodist Church Hall, 121 Otley Road, Harrogate. Last Friday each month.

Hastings and Battle Astronomical Society
Secretary: Mrs Karen Pankhurst, 20 High Bank Close, Ore, Hastings, E. Sussex.
Meetings: Details from Secretary.

Heart of England Astronomical Society
Secretary: Jean Poyner, 67 Ellerton Road, Kingstanding, Birmingham B44 0QE.
Meetings: Furnace End Village, every Thursday.

Hebden Bridge Literary & Scientific Society, Astronomical Section
Secretary: F. Parker, 48 Caldene Avenue, Mytholmroyd, Hebden Bridge, West Yorkshire.

Herschel Astronomy Society
Secretary: D. R. Whittaker, 149 Farnham Lane, Slough.
Meetings: Eton College, 2nd Friday each month.

Highlands Astronomical Society
Secretary: Richard Pearce, 1 Forsyth Street, Hopeman, Elgin.
Meetings: The Spectrum Centre, Inverness. 1st Tuesday each month, 7.30 p.m.

Howards Astronomy Club
Secretary: H. Ilett, 22 St Georges Avenue, Warblington, Havant, Hants.
Meetings: To be notified.

Huddersfield Astronomical and Philosophical Society
Secretary: R. A. Williams, 43 Oaklands Drive, Dalton, Huddersfield.
Meetings: 4A Railway Street, Huddersfield. Every Friday, 7.30 p.m.

Hull and East Riding Astronomical Society
Secretary: A. G. Scaife, 19 Beech Road, Elloughton, East Yorks.
Meetings: Wyke 6th Form College, Bricknell Avenue, Hull. 1st and 3rd Wednesday each month, Oct.–April, 7.30 p.m.

Ilkeston & District Astronomical Society
Secretary: Trevor Smith, 129 Heanor Road, Smalley, Derbyshire.
Meetings: The Friends Meeting Room, Ilkeston Museum, Ilkeston. 2nd Tuesday monthly, 7.30 p.m.

Ipswich, Orwell Astronomical Society
Secretary: R. Gooding, 168 Ashcroft Road, Ipswich.
Meetings: Orwell Park Observatory, Nacton, Ipswich. Wednesdays 8 p.m.

Irish Astronomical Association
Secretary: Michael Duffy, 26 Ballymurphy Road, Belfast, Northern Ireland.
Meetings: Room 315, Ashby Institute, Stranmills Road, Belfast. Fortnightly. Wednesdays, Sept.–April, 7.30 p.m.

Irish Astronomical Society
Secretary: c/o PO Box 2547, Dublin 15, Eire.

Isle of Man Astronomical Society
Secretary: James Martin, Ballaterson Farm, Peel, Isle of Man.
Meetings: The Manx Automobile Club, Hill Street, Douglas. 1st Thursday of each month, 8.30 p.m.

Isle of Wight Astronomical Society
Secretary: J. W. Feakins, 1 Hilltop Cottages, High Street, Freshwater, Isle of Wight.
Meetings: Unitarian Church Hall, Newport, Isle of Wight. Monthly.

Keele Astronomical Society
Secretary: Miss Caterina Callus, University of Keele, Keele, Staffs.
Meetings: As arranged during term time.

Kettering and District Astronomical Society
Asst. Secretary: Steve Williams, 120 Brickhill Road, Wellingborough, Northants.
Meetings: Quaker Meeting Hall, Northall Street, Kettering, Northants. 1st Tuesday each month. 7.45 p.m.

King's Lynn Amateur Astronomical Association
Secretary: P. Twynman, 17 Poplar Avenue, RAF Marham, King's Lynn.
Meetings: As arranged.

Lancaster and Morecambe Astronomical Society
Secretary: Miss E. Haygarth, 27 Coulston Road, Bowerham, Lancaster.
Meetings: Midland Hotel, Morecambe. 1st Wednesday each month except January. 7.30 p.m.

Lancaster University Astronomical Society
Secretary: c/o Students Union, Alexandra Square, University of Lancaster.
Meetings: As arranged.

Laymans Astronomical Society
Secretary: John Evans, 10 Arkwright Walk, The Meadows, Nottingham.
Meetings: The Popular, Bath Street, Ilkeston, Derbyshire. Monthly.

Leeds Astronomical Society
Secretary: A. J. Higgins, 23 Montagu Place, Leeds LS8 2RQ.
Meetings: Lecture Room, City Museum Library, The Headrow, Leeds.

Leicester Astronomical Society
Secretary: Ann Borell, 53 Warden's Walk, Leicester Forest East, Leics.
Meetings: Judgemeadow Community College, Marydene Drive, Evington, Leicester. 2nd and 4th Tuesdays each month, 7.30 p.m.

Letchworth and District Astronomical Society
Secretary: Eric Hutton, 14 Folly Close, Hitchin, Herts.
Meetings: As arranged.

Limerick Astronomy Club
Secretary: Tony O'Hanlon, 26 Ballycannon Heights, Meelick, Co. Clare, Ireland.
Meetings: Limerick Senior College, Limerick, Ireland. Monthly (except June and August), 8 p.m.

Lincoln Astronomical Society
Secretary: G. Winstanley, 36 Cambridge Drive, Washingborough, Lincoln.
Meetings: The Lecture Hall, off Westcliffe Street, Lincoln. 1st Tuesday each month.

Liverpool Astronomical Society
Secretary: David Whittle, 17 Sandy Lane, Tuebrook, Liverpool.
Meetings: City Museum, Liverpool. Wednesdays and Fridays, monthly.

Loughton Astronomical Society
Meetings: Loughton Hall, Rectory Lane, Loughton, Essex. Thursdays 8 p.m.

Lowestoft and Great Yarmouth Regional Astronomers (LYRA) Society
Secretary: R. Cheek, 7 The Glades, Lowestoft, Suffolk.
Meetings: Community Wing, Kirkley High School, Kirkley Run, Lowestoft. 3rd Thursday, Sept.–May. Afterwards in School Observatory. 7.15 p.m.

Luton & District Astronomical Society
Secretary: D. Childs, 6 Greenways, Stopsley, Luton.
Meetings: Luton College of Higher Education, Park Square, Luton. Second and last Friday each month, 7.30 p.m.

Lytham St Annes Astronomical Association
Secretary: K. J. Porter, 141 Blackpool Road, Ansdell, Lytham St Annes, Lancs.
Meetings: College of Further Education, Clifton Drive South, Lytham St Annes. 2nd Wednesday monthly Oct.–June.

Macclesfield Astronomical Society
Secretary: Mrs C. Moss, 27 Westminster Road, Macclesfield, Cheshire.
Meetings: The Planetarium, Jodrell Bank, 1st Tuesday each month.

Maidenhead Astronomical Society
Secretary: c/o Chairman, Peter Hunt, Hightrees, Holyport Road, Bray, Berks.
Meetings: Library. Monthly (except July) 1st Friday.

Maidstone Astronomical Society
Secretary: Stephen James, 4 The Cherry Orchard, Haddow, Tonbridge, Kent.
Meetings: Nettlestead Village Hall, 1st Tuesday in month except July and Aug. 7.30 p.m.

Manchester Astronomical Society
Secretary: J. H. Davidson, Godlee Observatory, UMIST, Sackville Street, Manchester 1.
Meetings: At the Observatory, Thursdays, 7.30–9 p.m.

Mansfield and Sutton Astronomical Society
Secretary: G. W. Shepherd, Sherwood Observatory, Coxmoor Road, Sutton-in-Ashfield, Notts.
Meetings: Sherwood Observatory, Coxmoor Road. Last Tuesday each month, 7.45 p.m.

Mexborough and Swinton Astronomical Society
Secretary: Mark R. Benton, 61 The Lea, Swinton, Mexborough, Yorks.
Meetings: Methodist Hall, Piccadilly Road, Swinton, Near Mexborough. Thursdays, 7 p.m.

Mid-Kent Astronomical Society
Secretary: Brian A. van de Peep, 11 Berber Road, Strood, Rochester, Kent.
Meetings: Medway Teachers Centre, Vicarage Road, Strood, Rochester, Kent. Last Friday in month. Mid Kent College, Horsted. 2nd Friday in month.

Milton Keynes Astronomical Society
Secretary: The Secretary, Milton Keynes Astronomical Society, Bradwell Abbey Field Centre, Bradwell, Milton Keynes MK1 39AP.
Meetings: Alternate Tuesdays.

Moray Astronomical Society
Secretary: Richard Pearce, 1 Forsyth Street, Hopeman, Elgin, Moray, Scotland.
Meetings: Village Hall Close, Co. Elgin.

Newbury Amateur Astronomical Society
Secretary: Mrs A. Davies, 11 Sedgfield Road, Greenham, Newbury, Berks.
Meetings: United Reform Church Hall, Cromwell Road, Newbury. Last Friday of month, Aug.–May.

Newcastle-on-Tyne Astronomical Society
Secretary: C. E. Willits, 24 Acomb Avenue, Seaton Delaval, Tyne and Wear.
Meetings: Zoology Lecture Theatre, Newcastle University. Monthly.

North Aston Space & Astronomical Club
Secretary: W. R. Chadburn, 14 Oakdale Road, North Aston, Sheffield.
Meetings: To be notified.

Northamptonshire Natural History Astronomical Society
Secretary: Dr Nick Hewitt, 4 Daimler Close, Northampton.
Meetings: Humphrey Rooms, Castillian Terrace, Northampton. 2nd and last Monday each month.

North Devon Astronomical Society
Secretary: P. G. Vickery, 12 Broad Park Crescent, Ilfracombe, North Devon.
Meetings: Pilton Community College, Chaddiford Lane, Barnstaple. 1st Wednesday each month, Sept.–May.

North Dorset Astronomical Society
Secretary: J. E. M. Coward, The Pharmacy, Stalbridge, Dorset.
Meetings: Charterhay, Stourton, Caundle, Dorset. 2nd Wednesday each month.

North Staffordshire Astronomical Society
Secretary: N. Oldham, 25 Linley Grove, Alsager, Stoke-on-Trent.
Meetings: 1st Wednesday of each month at Cartwright House, Broad Street, Hanley.

North Western Association of Variable Star Observers
Secretary: Jeremy Bullivant, 2 Beaminster Road, Heaton Mersey, Stockport, Cheshire.
Meetings: Four annually.

Norwich Astronomical Society
Secretary: Malcolm Jones, Tabor House, Norwich Road, Malbarton, Norwich.
Meetings: The Observatory, Colney Lane, Colney, Norwich. Every Friday, 7.30 p.m.

Nottingham Astronomical Society
Secretary: C. Brennan, 40 Swindon Close, Giltbrook, Nottingham.

Oldham Astronomical Society
Secretary: P. J. Collins, 25 Park Crescent, Chadderton, Oldham.
Meetings: Werneth Park Study Centre, Frederick Street, Oldham. Fortnightly, Friday.

Open University Astronomical Society
Secretary: Jim Lee, c/o above, Milton Keynes.
Meetings: Open University, Walton Hall, Milton Keynes. As arranged.

Orpington Astronomical Society
Secretary: Miss Lucinda Jones, 263 Crescent Drive, Petts Wood, Orpington, Kent BR5 1AY.
Meetings: Orpington Parish Church Hall, Bark Hart Road. Thursdays monthly, 7.30 p.m. Sept.–July.

Peterborough Astronomical Society
Secretary: Sheila Thorpe, 6 Cypress Close, Longthorpe, Peterborough.
Meetings: 1st Thursday every month at 7.30 p.m.

Plymouth Astronomical Society
Secretary: Sheila Evans, 40 Billington Close, Eggbuckland, Plymouth.
Meetings: Glynnis Kingdon Centre. 2nd Friday each month.

Port Talbot Astronomical Society (was Astronomical Society of Wales)
Secretary: J. A. Minopoli, 11 Tan Y Bryn Terrace, Penclowdd, Swansea.
Meetings: Port Talbot Arts Centre, 1st Tuesday each month, 7.15 p.m.

Portsmouth Astronomical Society
Secretary: G. B. Bryant, 81 Ringwood Road, Southsea.
Meetings: Monday. Fortnightly.

Preston & District Astronomical Society
Secretary: P. Sloane, 77 Ribby Road, Wrea Green, Kirkham, Preston, Lancs.
Meetings: Moor Park (Jeremiah Horrocks) Observatory, Preston. 2nd Wednesday, last Friday each month. 7.30 p.m.

The Pulsar Group
Secretary: Barry Smith, 157 Reridge Road, Blackburn, Lancs.
Meetings: Amateur Astronomy Centre, Clough Bank, Bacup Road, Todmorden, Lancs. 1st Thursday each month.

Reading Astronomical Society
Secretary: Mrs Muriel Wrigley, 516 Wokingham Road, Earley, Reading.
Meetings: St Peter's Church Hall, Church Road, Earley. Monthly (3rd Sat.), 7 p.m.

Renfrew District Astronomical Society (formerly Paisley A.S.)
Secretary: D. Bankhead, 3c School Wynd, Paisley.
Meetings: Coats Observatory, Oakshaw Street, Paisley. Fridays, 7.30 p.m.

Richmond & Kew Astronomical Society
Secretary: Stewart McLaughlin, 41A Bruce Road, Mitcham, Surrey.
Meetings: Richmond Central Reference Library, Richmond, Surrey.

Rower Astronomical Club
Secretary: Mary Kelly, Knockatore, The Rower, Thomastown, Co. Kilkenny, Eire.
Meetings: ?? ???.

Salford Astronomical Society
Secretary: J. A. Handford, 45 Burnside Avenue, Salford 6, Lancs.
Meetings: The Observatory, Chaseley Road, Salford.

Salisbury Astronomical Society
Secretary: Mrs R. Collins, Mountains, 3 Fairview Road, Salisbury, Wilts.
Meetings: Salisbury City Library, Market Place, Salisbury.

Sandbach Astronomical Society
Secretary: Phil Benson, 8 Gawsworth Drive, Sandbach, Cheshire.
Meetings: Sandbach School, as arranged.

Scarborough & District Astronomical Society
Secretary: Mrs S. Anderson, Basin House Farm, Sawdon, Scarborough, N. Yorks.
Meetings: Scarborough Public Library. Last Saturday each month, 7–9 p.m.

Scottish Astronomers Group
Secretary: G. Young c/o Mills Observatory, Balgay Park, Ancrum, Dundee.
Meetings: Bi-monthly, around the country. Syllabus given on request.

Sheffield Astronomical Society
Secretary: Mrs Lilian M. Keen, 21 Seagrave Drive, Gleadless, Sheffield.
Meetings: City Museum, Weston Park, 3rd Friday each month. 7.30 p.m.

Sidmouth and District Astronomical Society
Secretary: M. Grant, Salters Meadow, Sidmouth, Devon.
Meetings: Norman Lockyer Observatory, Salcombe Hill. 1st Monday in each month.

Society for Popular Astronomy (was Junior Astronomical Society)
Secretary: Guy Fennimore, 36 Fairway, Keyworth, Nottingham.
Meetings: Last Saturday in Jan., April, July, Oct., 2.30 p.m. in London.

Solent Amateur Astronomers
Secretary: R. Smith, 16 Lincoln Close, Woodley, Romsey, Hants.
Meetings: Room 2, Oaklands Community Centre, Fairisle Road, Lordshill, Southampton. 3rd Tuesday.

Southampton Astronomical Society
Secretary: M. R. Hobbs, 124 Winchester Road, Southampton.
Meetings: Room 148, Murray Building, Southampton University, 2nd Thursday each month, 7.30 p.m.

South Downs Astronomical Society
Secretary: J. Green, 46 Central Avenue, Bognor Regis, West Sussex.
Meetings: Assembly Rooms, Chichester. 1st Friday in each month.

South-East Essex Astronomical Society
Secretary: C. Jones, 92 Long Riding, Basildon, Essex.
Meetings: Lecture Theatre, Central Library, Victoria Avenue, Southend-on-Sea. Generally 1st Thursday in month, Sept.–May.

South-East Kent Astronomical Society
Secretary: P. Andrew, 7 Farncombe Way, Whitfield, nr. Dover.
Meetings: Monthly.

South Lincolnshire Astronomical & Geophysical Society
Secretary: Ian Farley, 12 West Road, Bourne, Lincs.
Meetings: South Holland Centre, Spalding. 3rd Thursday each month, Sept.–May. 7.30 p.m.

South London Astronomical Society
Chairman: P. Bruce, 2 Constance Road, West Croydon CR0 2RS.
Meetings: Surrey Halls, Birfield Road, Stockwell, London SW4. 2nd Tuesday each month, 8 p.m.

Southport Astronomical Society
Secretary: R. Rawlinson, 188 Haig Avenue, Southport, Merseyside.
Meetings: Monthly Sept.–May, plus observing sessions.

Southport, Ormskirk and District Astronomical Society
Secretary: J. T. Harrison, 92 Cottage Lane, Ormskirk, Lancs L39 3NJ.
Meetings: Saturday evenings, monthly as arranged.

South Shields Astronomical Society
Secretary: c/o South Tyneside College, St George's Avenue, South Shields.
Meetings: Marine and Technical College. Each Thursday, 7.30 p.m.

South Somerset Astronomical Society
Secretary: G. McNelly, 11 Laxton Close, Taunton, Somerset.
Meetings: Victoria Inn, Skittle Alley, East Reach, Taunton. Last Saturday each month, 7.30 p.m.

South-West Cotswolds Astronomical Society
Secretary: C. R. Wiles, Old Castle House, The Triangle, Malmesbury, Wilts.
Meetings: 2nd Friday each month, 8 p.m. (Sept.–June).

South-West Herts Astronomical Society
Secretary: Frank Phillips, 54 Highfield Way, Rickmansworth, Herts.
Meetings: Rickmansworth. Last Friday each month, Sept.–May.

Stafford and District Astronomical Society
Secretary: Mrs L. Hodkinson, Beecholme, Francis Green Lane, Penkridge, Staffs.
Meetings: Riverside Centre, Stafford. Every 3rd Thursday, Sept.–May, 7.30 p.m.

Stirling Astronomical Society
Secretary: Mrs C. Traynor, 5c St Mary's Wynd, Stirling.
Meetings: Smith Museum & Art Gallery, Dumbarton Road, Stirling. 2nd Friday each month, 7.30 p.m.

Stoke-on-Trent Astronomical Society
Secretary: M. Pace, Sundale, Dunnocksfold Road, Alsager, Stoke-on-Trent.
Meetings: Cartwright House, Broad Street, Hanley. Monthly.

Sussex Astronomical Society
Secretary: Mrs C. G. Sutton, 75 Vale Road, Portslade, Sussex.
Meetings: English Language Centre, Third Avenue, Hove. Every Wednesday, 7.30–9.30 p.m. Sept.–May.

Swansea Astronomical Society
Secretary: D. F. Tovey, 43 Cecil Road, Gowerton, Swansea.
Meetings: Dillwyn Llewellyn School, John Street, Cockett, Swansea. 2nd and 4th Thursday each month at 7.30 p.m.

Tavistock Astronomical Society
Secretary: D. S. Gibbs, Lanherne, Chollacott Lane, Whitchurch, Tavistock, Devon.
Meetings: Science Laboratory, Kelly College, Tavistock. 1st Wednesday in month. 7.30 p.m.

Thames Valley Astronomical Group
Secretary: K. J. Pallet, 82a Tennyson Street, South Lambeth, London SW8 3TH.
Meetings: Irregular.

Thanet Amateur Astronomical Society
Secretary: P. F. Jordan, 85 Crescent Road, Ramsgate.
Meetings: Hilderstone House, Broadstairs, Kent. Monthly.

Torbay Astronomical Society
Secretary: R. Jones, St Helens, Hermose Road, Teignmouth, Devon.
Meetings: Town Hall, Torquay. 3rd Thursday, Oct.–May.

Tullamore Astronomical Society
Secretary: S. McKenna, 145 Arden Vale, Tullamore, Co. Offaly, Eire.
Meetings: Tullamore Vocational School, Fortnightly, Tuesdays, Oct–June. 8 p.m.

Tyrone Astronomical Society
Secretary: John Ryan, 105 Coolnafranky Park, Cookstown, Co. Tyrone.
Meetings: Contact Secretary.

Usk Astronomical Society
Secretary: D. J. T. Thomas, 20 Maryport Street, Usk, Gwent.
Meetings: Usk Adult Education Centre, Maryport Street. Weekly, Thursdays (term dates).

Vectis Astronomical Society
Secretary: J. W. Smith, 27 Forest Road, Winford, Sandown, I.W.
Meetings: 4th Friday each month, except Dec. at Lord Louis Library Meeting Room, Newport, I.W.

Vigo Astronomical Society
Secretary: Robert Wilson, 43 Admers Wood, Vigo Village, Meopham, Kent.
Meetings: Vigo Village Hall, as arranged.

Webb Society
Secretary: M. B. Swan, 194 Foundry Lane, Freemantle, Southampton, Hants.
Meetings: As arranged.

Wellingborough District Astronomical Society
Secretary: S. M. Williams, 120 Brickhill Road, Wellingborough, Northants.
Meetings: On 2nd Wednesday. Gloucester Hall, Church Street, Wellingborough, 7.30 p.m.

Wessex Astronomical Society
Secretary: Leslie Fry, 14 Hanhum Road, Corfe Mullen, Dorset.
Meetings: Allendale Centre, Wimborne, Dorset. 1st Tuesday of each month.

West of London Astronomical Society
Secretary: Tom. H. Ella, 25 Boxtree Road, Harrow Weald, Harrow, Middlesex.
Meetings: Monthly, alternately at Hillingdon and North Harrow. 2nd Monday of the month, except August.

West Midlands Astronomical Association
Secretary: Miss S. Bundy, 93 Greenridge Road, Handsworth Wood, Birmingham.
Meetings: Dr Johnson House, Bull Street, Birmingham. As arranged.

West Yorkshire Astronomical Society
Secretary: K. Willoughby, 11 Hardisty Drive, Pontefract, Yorks.
Meetings: Rosse Observatory, Carleton Community Centre, Carleton Road, Pontefract, each Tuesday, 7.15 to 9 p.m.

Whittington Astronomical Society
Secretary: Peter Williamson, The Observatory, Top Street, Whittington, Shropshire.
Meetings: The Observatory every month.

Wolverhampton Astronomical Society
Secretary: M. Astley, Garwick, 8 Holme Mill, Fordhouses, Wolverhampton.
Meetings: Beckminster Methodist Church Hall, Birches Road, Wolverhampton. Alternate Mondays, Sept.–April.

Worcester Astronomical Society
Secretary: Arthur Wilkinson, 179 Henwick Road, St Johns, Worcester.
Meetings: Room 117, Worcester College of Higher Education, Henwick Grove, Worcester. 2nd Thursday each month.

Worthing Astronomical Society
Contact: G. Boots, 101 Ardingly Drive, Worthing, Sussex.
Meetings: Adult Education Centre, Union Place, Worthing, Sussex. 1st Wednesday each month (except August). 7.30 p.m.

Wycombe Astronomical Society
Secretary: P. A. Hodgins, 50 Copners Drive, Holmer Green, High Wycombe, Bucks.
Meetings: 3rd Wednesday each month, 7.45 p.m.

York Astronomical Society
Secretary: Simon Howard, 20 Manor Drive South, Acomb, York.
Meetings: Goddricke College, York University. 1st and 3rd Fridays.

Any society wishing to be included in this list of local societies or to update details are invited to write to the Editor (c/o Macmillan Reference, 25 Eccleston Place, London SW1W 9NF), so that the relevant information may be included in the next edition of the *Yearbook*.